T5-BCI-283

King of Colorado Botany

King of Colorado Botany

Charles Christopher Parry, 1823–1890

William A. Weber

UNIVERSITY PRESS OF COLORADO

Published by the University Press of Colorado
P.O. Box 849
Niwot, Colorado 80544
(303)530-5337

Printed in the United States of America.

The University Press of Colorado is a cooperative publishing enterprise supported, in part, by Adams State College, Colorado State University, Fort Lewis College, Mesa State College, Metropolitan State College of Denver, University of Colorado, University of Northern Colorado, University of Southern Colorado, and Western State College of Colorado.

The paper used in this publication meets the minimum requirements of the American National Standard for Information Sciences — Permanence of Paper for Printed Library Materials. ANSI Z39.48-1984

Library of Congress Cataloging-in-Publication Data

Weber, William A. (William Alfred), 1918-
King of Colorado botany : Charles Christopher Parry, 1823-1890 / William A. Weber.
p. cm.
Includes bibliographical references (p.) and index.
ISBN 0-87081-431-1 (cloth : alk. paper)
1. Parry, Charles Christopher, 1823-1890. 2. Botanists--Colorado--Biography. 3. Botany--Colorado--History. 4. Parry, Charles -Description and travel. I. Title
QK31.P35W43 1997
580'.92--dc21
[B] 97-1544
CIP

10 9 8 7 6 5 4 3 2 1

Contents

Preface

This project began with my offer to identify for a colleague at Duke University a large collection of plants collected along the Colorado Front Range in 1862 by Charles C. Parry, Elihu Hall, and J. P. Harbour. Because this was a historic collection for which a list was published in 1863 by Asa Gray, I decided to compile a concordance to see how taxonomic concepts had changed over the past 130 years, and to provide an annotated list with modern nomenclature. The work expanded with my examination of Parry's own herbarium at Iowa State University. The Parry, Hall, and Harbour collections are duplicated in many herbaria throughout the United States and Europe, and for reasons to be explained, many of these duplicate specimens have never been given names, and some have never been labeled.

The specimen labels themselves lack locality and ecological information; for this reason, except for those from which new species were named (type specimens), the specimens have been largely ignored by botanists. Itineraries of the various expeditions were thought to be lacking. Scientists criticize the brevity of Parry's labels, (see p. 84) and one is hard pressed to defend his lack of locality data. However, in the early days of exploration, most collectors never realized that precise locality data might someday become important, especially where type specimens were concerned; moreover, named features were not available on the maps of the time, and many settlements were ephemeral. Nevertheless, it does seem lax for Parry not to have mentioned South Park, or some of the mountains he himself named, as well as the actual dates of collection, so that reference to an itinerary could be helpful.

The mystery of the label brevity may have to do with the fact that the collection, except for unicates collected by Parry alone, became what is known as the "Hall & Harbour collections." That Parry was associated with them was largely overlooked or disbelieved. The very large collections were worked up for wide distribution and sale by Hall & Harbour. Many sets were sold and distributed before a master set was identified by Asa Gray. Parry was not given credit on the labels for his participation in the expedition. However, in researching the literature of the time, I found that Parry had, in fact, published detailed accounts of the regions that he and his comrades explored, and that an itinerary of the most important expedition of 1862 was published in a little-known paper by George Engelmann. These are the first in-depth field reports written by an acute scientific observer and passionate mountain enthusiast since the journal of Edwin James with the Long Expedition in 1820. James, however, traveled as part of a military party; he worked within the constraints of its itinerary and, with the exception of his ascent of Pike's Peak, visited only the outer foothills of the Front Range of Colorado. Parry, Hall, and Harbour went on their own, with the sole mission of collecting the flora and distributing duplicate sets far and wide.

Parry had greater goals in mind. He was caught up in the romance of U.S. expansion westward and believed, like so many of his contemporaries, in manifest destiny. On his travels, Parry constantly looked at the landscape in terms of its ability to sustain agriculture and human populations. He was always looking for a feasible railroad route across the plains and through the Rocky Mountains to the Pacific Slope. He was a one-man "chamber of commerce," and in a number of newspaper articles published in Chicago, he was obviously trying to entice easterners to go to Colorado and build a "mountain empire." Parry, in fact, might be said to have been the first resident botanist of Colorado, in that for almost two decades he maintained a summer cabin at the foot of Gray's and Torrey's Peaks.

Parry meant more to the world's understanding of the botany of Colorado than any other person of the early exploratory period. Yet, except for his name having been affixed to the name of many of the state's wildflowers, two mountain peaks, and a creek in Middle Park, as a person he is hardly known in Colorado. In combining the lists of Parry's collections with the scientific and semipopular descriptions of his travels—until now only to be found scattered in rare periodicals and old newspapers—I hope to bring his contribution into proper focus.

This memoir deals only with the Colorado period of Parry's career. Extensive references to other aspects, especially his relationships with other field botanists, are to be found in several books on western American botanical history (Dupree 1959; Lenz 1986; McVaugh 1956; Rodgers 1942, 1944). A large collection of Parry's correspondence and other papers is filed in the Iowa State University Library (McMullen 1995).

Acknowledgments

I am greatly indebted to Deborah Q. Lewis, curator, for the loan of the Parry plant collections at the Ada Hayden Herbarium of Iowa State University at Ames and for her research in locating important parts of the Parry archive at the university library, and to Betty Erickson, of the Iowa State University Library Special Collections, for permitting use of much of the material (MS-290) used herein. I am also grateful for the kindness of Dr. Robert Wilbur, of Duke University, for the loan of a large set of the Parry, Hall, and Harbour collections under his curation there. I am indebted to Iowa State University, which I attended as an undergraduate and as Prof. Ada Hayden's assistant in 1938–1940, and where I first became acquainted with the Parry herbarium. I was the photostat operator at the university library in 1946 and discovered that Parry's notebooks of barometric readings, some of which have been very useful to me over the years, are housed there. Science is fortunate to have found caretakers of such valuable historic materials. Two photographs of Parry are included in this volume, courtesy of the Hunt Institute for Botanical Documentation, Carnegie Mellon University, Pittsburgh, PA. The photographs of mountain peaks named by Parry were kindly provided by Dr. David Hill of Boulder, CO, an expert in electromagnetic theory as well as mountaineering.

Acknowledgments

King of Colorado Botany

Introduction

The name Charles Christopher Parry (1823–1890) is well known to botanists worldwide. Leaving a very brief medical practice in Davenport, IA (1846–1849), Parry joined the Mexican Boundary Survey (1849–1852) as assistant surgeon; botanized with Edward Palmer in Mexico (including Baja California) (1877–1878), Utah, and California; and joined a reconnaissance in Wyoming (1873) with Capt. William A. Jones. Parry was an avid botanical explorer, and many species of plants bear his name. He is not to be confused with Sir William Edward Parry, 1790–1855, the English navigator and Arctic explorer.

Parry's greatest contribution to science is probably his exploration of the southern Rocky Mountains of Colorado; through the distribution of his botanical collections he introduced the Colorado flora to the world. He was also fascinated with mountains, and he made barometric observations that permitted the first accurate estimations of the altitudes of the high peaks of Colorado. Thanks to Parry, the region of Upper Clear Creek commemorates several great botanical personalities of his time. The ridge of the Continental Divide north from Empire has as one anchor James Peak—for Edwin James, botanist with the Long Expedition, named by Parry (cf. Farquhar 1961)—followed in succession southward by Parry's Peak, 13,391 feet, Mount Eva (commemorating Parry's wife), and the Mount Flora of Parry's accounts. Parry's Creek leads northwestward down from Parry's Peak to Middle Park and the present village of Winter Park. Mount Engelmann, with its majestic cirque, lies to the west of Empire, and Gray's and Torrey's Peaks, to the south, stand high above the valley of Clear Creek. A second Parry's Peak, 12,682 feet, lies just south of Mount Elbert and the village of Twin Lakes, in Lake County (Mount Elbert Quadrangle), although Parry is not known to have visited the area.

Over eighty new species of flowering plants were named from Parry's Colorado collections, and many more from his collections in Mexico and the U.S. Southwest. Two genera, *Neoparrya* and *Parryella,* were dedicated to him. Some of the Colorado novelties, such as *Astragalus parryi, Campanula parryi, Chrysothamnus parryi, Helianthella parryi, Oxytropis parryi, Pedicularis parryi, Primula parryi,* and *Trifolium parryi* are Colorado's most well-known wildflowers. Sir Joseph Hooker, the great plant geographer and director of the Royal Botanic Gardens at Kew, called Parry "King of Colorado Botany." Parry's plant collections were distributed in duplicate to a great many institutions: Harvard University, New York Botanical Garden, Philadelphia Academy of Sciences, Missouri Botanical Garden, Smithsonian Institution, Royal Botanic Gardens (Kew), Oxford University, Edinburgh Botanic Garden, British Museum, as well as botanical gardens in Brussels, Florence, Geneva, Manchester, and St. Petersburg. Parry thus introduced the world to the wonders of the vegetation of the Colorado Rocky Mountains.

Beyond these honorifics, however, Parry remains less known in Colorado and the United States than in the rest of the world. In the first volume of the *Flora of North America* (1993), the chapter on history entitled "Taxonomic Botany and Floristics" fails to mention Parry except for two brief name references. His Rocky Mountain work is not discussed. One reason for this obscurity is that during Parry's lifetime there was not an established herbarium in the entire Rocky Mountain region for a set of collections to be housed and cared for. All of Parry's collections in the United States were confined to a few great herbaria of the eastern part of the country. In order to write a flora of Colorado, one had three alternatives: travel to these places to study the specimens; borrow them in small lots; or try, through one's own field work, to rediscover the species that were still poorly known. Harold D. Harrington was brought to Colorado State College with a mission—to produce a *Flora of Colorado* (Harrington 1954), but he never had the financial support to travel to the herbaria that housed the historical collections. Hence he was never able to verify the identity or establish the validity of many reported species.

A fair portion of the flora of Colorado was well known long before any botanist set foot inside the state. Some wide-ranging species, mostly of the high plains, were discovered on Lewis and Clark's travels (1804–1806) and by Thomas Nuttall (1811–1812). The first and justly famous expedition within Colorado was the Long Expedition (1820), with Edwin James as botanist (Goodman and Lawson 1995). This expedition skirted the eastern base of the Rocky Mountains, entering the area along the South Platte River to the present site of Denver, then moving southward to Pike's Peak and the Arkansas River, and leaving by way of the Mesa De Maya south of present-day Rocky Ford. James's most notable accomplishment on this expedition was his climb and botanical sampling of Pike's Peak, on which were recorded the first alpine species to be found in Colorado. A lull followed this historical beginning of Colorado floristic botany. Although John Charles Frémont's expeditions increased interest in the Rocky Mountains, few of his Colorado collections survived. Prior to Parry's travels, all botanical exploration of Colorado was supported by U.S. Army escorts on military and pathfinding missions. This lull ended with the arrival on the Colorado scene of Charles Christopher Parry.

The Collections

Parry visited Colorado several times, beginning in 1861. His most important plant collection was that of 1862, unfortunately known forever after as the Hall & Harbour collections. Elihu Hall and J. P. Harbour, both from Illinois—and neither of whom, it appears, had previously ever been to Colorado or had any knowledge of the Rocky Mountain flora—joined Parry in an expedition to the Colorado Front Range and South Park. Parry acted as guide and botanical expert. However, the specimens were eventually provided with labels that not only lacked locality data and dates, but omitted, most likely for reasons of modesty, the name of Parry as a joint collector. Parry sent a set of this collection to Harvard University, where Asa Gray evaluated the specimens and published the results, as he had for Parry's 1861 collections. Hall sold duplicate sets to herbaria in the United States and Europe. Many species were described for the first time from this collection, but because of the lack of collection data, most of the remaining specimens are of little value except for their representation of the species. Unfortunately, although field notebooks covering 1848 through 1852 are listed in the inventory of Parry papers (McMullen 1995), none are listed for the most important years (1861–1866).

In July 1992 Dr. Robert Wilbur, curator of the herbarium at Duke University, loaned me a collection from the Hall & Harbour collections of 1862 that he had received some years ago from another institution. These specimens had never been unpacked and bore no identifications. I offered to name them, and this provided a unique opportunity for me to examine a very large (although not quite complete) set of this collection. I decided to take this chance to compare my identifications with the names provided by Asa Gray (Gray 1863). As Gray said, "The plants were numbered and distributed into sets by Messrs. Hall and Harbour before they were seen by me, and a full set was supplied to me for examination, which serves as a basis for the following list. This accounts for a few misplacements, and also for the occasional mixture of two species under the same number, which, under the circumstances, it was not easy altogether to avoid. The collectors appear to have been somewhat too fearful of distributing the same species under two or more numbers; but the opposite course, in case of doubt, is preferable. Even well-marked varieties had better be kept separate in distributed collections."

About half the specimens in the DUKE set had labels, but few had names beyond the genus; the rest had no labels, but only penciled numbers. However, because the numbers corresponded with those in Gray's published list, I was able to prepare labels for them. The set was large but incomplete. A more complete intact set exists in the Parry Herbarium at Ames, IA. I examined all of the material, made comparisons with the list published by Gray in 1863, and brought the nomenclature up to date. To attempt to locate the equivalent collections at the Gray Herbarium at Harvard University, where they are dispersed among millions, seemed to me not to be worth the effort, considering that most of the specimens are not critical taxonomically and the holotypes are being catalogued actively by the staff at the Gray.

With few exceptions, the Hall & Harbour collections at Ames, as well as Parry's own collection there, have never been examined. A concordance of all of the Colorado collections would not only improve the utility of the DUKE and ISC collections, but would benefit the other herbaria containing duplicate material that probably had similarly inadequate data. At the same time, it is appropriate to bring together some of Parry's published descriptions of his itinerary and impressions of Colorado in the 1860s.

The Life of C. C. Parry

Professional botanists should consult the best account of Parry's career, presented by Joseph A. Ewan (1950, 1981). Ewan's work points out the excellence of Parry's work as an explorer, Parry's ability to write fine prose and circulate his observations widely to the general public through the newspapers, and his lack of attention to detail regarding the provenance of his collections.

The following is probably the best biographical sketch of Parry, written by a close colleague (Preston 1893).

On the twentieth of February, 1890, there died at his pleasant home near this city, one to whom the Davenport Academy of Sciences was deeply indebted, and whose memory, fragrant and pure as the flowers he loved, it will ever cherish.

At that time the working force of our Association had been almost paralyzed by recent sad losses; the publication of its Proceedings was for a time deferred, and so it happened that he who was always ready with an appreciative tribute to the memory of associate or friend, has waited thus long for an expression from this, his home Society, of the admiration and esteem which each and all of its members entertained for him. It is to be regretted that there was not found among us some co-laborer in his own field of botanical science to prepare a sketch of Dr. Parry's life and work—an undertaking for which the writer is qualified only by warm personal friendship and long association in the affairs of the Academy. Deficiencies which must in consequence of necessity exist will, however, in part be made good by citations from those better qualified to speak.

Charles Christopher Parry was born in the hamlet of Admington, Gloucestershire, England, August 28, 1823. Descended through a long line of clergymen of the Established Church, he was himself of a deeply religious nature, and rarely endowed with that poetic feeling and insight so apt to characterize the true naturalist.

In 1832 the family removed to America, settling on a farm in Washington County, New York. Here the remainder of his boyhood was passed, and, the advantages of the schools of the locality having been well improved, he entered Union College at Schenectady, and in due time was graduated therefrom with honors. He began the study of medical botany in his undergraduate years, and subsequently received the degree of Doctor of Medicine from Columbia College. [There is no mention anywhere of Parry having met John Torrey while at Columbia, but this can be inferred by the close contact that existed between the two in Parry's subsequent botanical collections in Iowa; cf. Rodgers (1942).]

Coming West and to Davenport in the fall of 1846, he entered upon the practice of his profession, but continued in it for a few months only, very soon discovering that all his natural tastes and instincts led directly away from the unreason, the too-often self-inflicted ills, and the petty conflicts with which the active physician has perforce to deal—led him to the unvexed, blossoming solitudes where Nature, silent and orderly, works out her fair results.

His earliest collecting had been done in the attractive floral region about his home in northeastern New York, in the summer of 1842 and the four years following; and now again, attracted to this more congenial work, we find him employing much of the season of 1847 in making a collection of the wild flowers about Davenport, of which, with the dates of finding, he has left a manuscript list. Those of us who knew him well in after years can readily picture the brisk, dark-complexioned, though blue-eyed youth, symmetrically but slightly built and somewhat below the medium height, in his solitary quest by riverside and deep ravine, over wooded bluff and prairie expanse, for the treasures which were more to him than gold—for such early friends as the prairie primrose, the moccasin-flower, and the gentian, which in later years he complained had been quite driven out by the blue-grass and white clover.

In the course of that summer, also, he accompanied a United States surveying party, under Lieutenant J. Morehead, on an excursion into Central Iowa, in the vicinity of the present State capital. From this time on (except for a short time while connected with the Mexican Boundary Survey, when he discharged the duties of Assistant Surgeon), the physician was merged into the naturalist. He was almost continuously in the field collecting, but Davenport remained his home. Here, in 1853, he was married to Miss Sarah M. Dalzell, who, dying five years later, left him with an only child, a daughter. But she, too, a fair, unfolding flower, was claimed by death at an early age.

In 1859, he was married again, to Mrs. E. R. Preston of Westford, Connecticut, who, through the more than thirty years of their union, entered helpfully into all his work and plans, assisting him in his study and often accompanying him to the field, and who is left to mourn the loss of one who, in every relation of life, was exceptionally unselfish and kind. Of his two brothers and six sisters only two remain, viz.: Joseph Parry and Mrs. Charles Pickering, both of Davenport, beside a half-sister, Mrs. Austin, residing in Arkansas.

We are fortunate in possessing, in Dr. Parry's own words (Proc., Vol. II, p. 279), a succinct, chronological account of his work up to 1878, which need not be repeated here. Suffice it to say that for more than thirty years the greater part of his time had been spent in observing and collecting—the St. Peters and up the St. Croix; across the Isthmus to San Diego, to the junction of the Gila and Colorado, along the Southern boundary line and up the coast as far as Monterey; through Texas to El Paso, to the Pima settlements on the Gila, and along the Rio Grande; in the mountains of Colorado, to which and to those of California he returned again and again in the pursuit of his special study, the Alpine Flora of North America; across the continent with a Pacific railroad surveying party by way of the Sangre de Cristo Pass, through New Mexico and Arizona, through the Tehachapi Pass, through the Tulare and San Joaquin Valleys to San Francisco; through the Wind River district to the Yellowstone National Park; in the valley of the Virgen and about Mount Nebo, Utah; about San Bernardino, California, and in the arid regions stretching to the eastward; and in New Mexico about San Luis Potosí, Saltillo, and Monterey.

The winter of 1852 was spent in Washington, in the preparation of his report as Botanist to the Mexican Boundary Survey; and the years from 1869 to 1871 inclusive, while Botanist to the United States Agricultural Department, were also passed chiefly at the capital, employed in arranging the extensive botanical collections from various government explorations, which had accumulated at the Smithsonian Institution. During this period, also, he visited, in his official capacity, the Royal Gardens and herbaria at Kew, England, and was attached as Botanist to the Commission of Inquiry which visited San Domingo early in 1871. The report of his observations on that island is a valuable summary of its chief botanical features, vegetable products, and agricultural capacities.

His visit to Kew and the land of his birth was the beginning of a lasting friendship between himself and the eminent Sir Joseph Hooker, Director of the Gardens, who afterward in a congratulatory letter dated February 27, 1877, calls him "already king of Colorado botany," and expressed deep interest in the results of his explorations, then making, in Southern California.

Subsequent to 1878, the date of the autobiographical sketch before mentioned, his work, although arduous and important, may be briefly summed up as follows:

In 1879, being called to the East by illness and death of his father, he did little if any work in the field. In 1880, as special agent of the Forestry Department of the United States Census Office, he accompanied Dr. Engelmann and Professor Sargent in an expedition to the Valley of the Columbia and the far Northwest. Wintering in California he spent the following year in that State, making numerous collecting trips North and South, including a trip to the Yosemite in June. Home again, in the summer of 1882, he was busily employed for some months in arranging his collections and in work for our Academy Proceedings. In the fall of that year he returned to California, and passed the winter in San Diego.

In January and February, 1883, he made two camping trips into Lower California; then, going to San Francisco, made numerous excursions from that point, and returned to Davenport in September. In June, 1884, he sailed a second time to England, returning in August of the following year, after spending much time at Kew, and visiting other herbaria and gardens on the Continent.

The summer of 1886 he spent partly with friends in Wisconsin, partly in the quiet enjoyment of his Iowa home. But even when resting, his mind did not rest—his wonderfully voluminous correspondence went on, and the microscope filled in his otherwise leisure hours. Again the winter was passed in San Francisco, from which city he made numerous collecting trips as before. Remaining in California, chiefly in the vicinity of San Francisco, until September, 1888, he was busily employed making special collections of *Arctostaphylos* and *Ceanothus*, and in the study of these and the genus *Alnus*. His last visit to California was made in the spring of 1889. Returning to Davenport in July, he made a trip to Canada and New England, visited New York and Philadelphia, and returned to his home but a few weeks before his death.

Most intimately connected with the botany of the Pacific Coast; "treading reverently in the steps of Chamisso, Douglas, Nuttall, and others of less note," who, at such accessible points as San Diego, Santa Barbara, Monterey, and the mouth of the Columbia, had, at an early day, preceded him, he greatly extended their labors. "None of the early investigators," says a writer in the *Century Magazine* (Oct., 1892), "was more typical than the late Dr. C. C. Parry, who first crossed the country with the Mexican Boundary Commission. At intervals, for forty years after, he was a familiar figure to hunters, prospectors, mountaineers, and all sorts of outdoor people, from the Arizona deserts to the Siskiyou pine forests."

Dr. Parry was recognized as an authority by botanists everywhere; not only in this country (where he ranked with the first) and in England, but on the Continent as well; and this notwithstanding the fact that he never published a book, had no ambition in the way of authorship, and left most of his discoveries to be described by others. His writings, though sufficient to constitute volumes, and comprising much of scientific value, are scattered in fragmentary form through various government and society reports, scientific journals, and the daily press. A list, approximately complete, will be published in connection herewith.

In 1875 he was made a fellow of the American Association for the Advancement of Science, in which body his membership dates back to 1851. He kept up a corresponding membership in the Philadelphia, Buffalo, St. Louis, Chicago, and California Academies of Science, and was connected with various other organizations, among them the Philosophical Society of Washington, D.C., the Bay District Horticultural Society of California, and the State Historical Society of Iowa. Of our own Academy he was, from the start, a most active promoter and one of the main supports. Its welfare was a matter of constant solicitude with him, and to his valuable papers, published in our Proceedings, the Academy's favorable recognition abroad is in great part due. Although absent in Arizona at the time of its organization, he was made a member of the first Board of Trustees, and continued in that capacity as long as he lived. On the resignation of our first President, Prof. Sheldon, in 1868, Dr. Parry was chosen to succeed him, and reelected again and again, until, in 1875, he declined longer to retain a place from which, and its duties, he must of necessity be much of the time absent. As a member of the Publication Committee from its inception, his counsel and assistance were invaluable, as indeed they were, while he lived, in the Academy's every undertaking.

Wholly free from that jealous self-seeking which too often mars genuine merit, his relations with his fellow-workers, whether tyros or masters in the science, were always of the pleasantest. The veteran botanist, Prof. John Torrey of Columbia college, to whose assistance and encouragement, from the time of their first acquaintance in 1845, he acknowledged himself deeply indebted, was his warm personal friend through life. Of their last living interview, which occurred in September, 1872, shortly before Torrey's death "full of years and honors," Dr. Parry writes in an obituary notice prepared for this Academy: "It was my privilege to entertain this distinguished guest at my rude botanical retreat in the heart of the Rocky Mountains. Here, in close proximity to my cabin, I could point out to him many of the living plants that he had described fifty years previously from herbarium specimens, but had never before seen in their living beauty." Owing to the early severity of the season at the time of this visit, Dr. Torrey was prevented from making the ascent of the peak to which his name had been given by his host and friend, although permitted "to gaze on its sky-piercing summit and to snatch from its wintry slopes some late-grown floral mementos of his early labors." Of this and its companion peak, Mt. Gray, Dr. Parry says: "In my first botanical exploration of the Rocky Mountain Region of Colorado, in 1861, I applied the name of Torrey and Gray to twin peaks which, from a distant view, had often attracted my attention. In the year following I succeeded in reaching the summit of the eastern peak, now well known as Gray's Peak, and determined its elevation by barometric observation. Two years afterward, in 1872, I stood for the second time on the same elevation, accompanied by Prof. Gray himself and a large party of acquaintances. Prof. Gray, pointing to the closely-adjoining western peak, expressed the earnest wish, seconded by all present, that it should continue to bear the name first affixed—of Mount Torrey—in worthy commemoration of his early and valued scientific associate."

It was Dr. Parry's pleasant privilege also to give its name to Mt. Guyot, in honor of his friend, Prof. Arnold Guyot of Princeton. His own name (bestowed by Surveyor-General F. M. Case) is borne by a peak of the Snowy Range to the north-west of Empire City. Farther removed from the abodes of men, retiring yet not inconspicuous, it stands amongst its fellows, an enduring and a fitting monument to him whom his friends knew as "good Doctor Parry."

Not less close than with Torrey and Gray were his relations with Dr. George Engelmann of St. Louis, whose death occurred in 1885. "Since my first acquaintance with him, in 1848," he writes, "when I called on him at St. Louis before starting on my first exploring trip with Dr. D. C. Owen in the then Northwest, our friendly intercourse has been constant, and the letters received from him would make up a respectable volume. How much I owe to his wise counsels, his substantial encouragement, and not less to his sharp criticisms (always well-meant), I can now best realize by feeling their loss. He knew just what to look for, and, when seen, he also knew its significance in elucidating the system of nature." This was not less true of Dr. Parry himself. Torrey, Gray, Engelmann, Parry! What were American botany but for these four co-laborers whose work and fame are inseparably interlinked?

Dr. Parry was essentially a field student, and the general accuracy of his conclusions is largely due to the fact that his observations were all made at first-hand; to this and to the thoroughness of his determinations, which were based on careful dissections of all accessible fruit, as well as of the flowering specimen, so that he was generally able, as he declared, to discriminate species by the fruit alone.

Industrious and indefatigable, "the bulk and value of his collections have probably not been equalled in America." (I quote from the Bulletin of the Torrey Botanical Club.) Beside contributing largely to the collections of his botanical friends and of various societies at home and abroad, he made for himself one of the finest private herbaria in the land, a collection, systematically classified and arranged, comprising over 18,000 determined specimens representative of nearly 6,800 species, together with some 1,400 specimens determined only as far as the genus. But while himself thus chiefly occupied in collecting from untrodden heights and tangled wilds, he recognized "with respect and reverence" the magnitude of the task assumed "by those masters of botanical science who have taken upon their broad shoulders the burden of a systematic arrangement of the whole vegetable kingdom."

Appreciating the beautiful as he did wherever found, and especially embodied in floral and arboreal forms, Dr. Parry was yet, for a naturalist, markedly utilitarian. Wherever he went, in whatever he did, his eyes were open to the practical. The plant, the tree which gave promise of usefulness was to him doubly interesting, and he spared no pains to obtain for such the recognition they deserved. To bring the Mexican rose into cultivation, for example, he made an extra trip into Lower California. He was at especial pains to introduce the remarkable *Spiraea caespitosa* or *tree-moss* found in the Wasatch Mountains, of which he writes: "The peculiar adaptation of this plant for ornamental rock-work can be appreciated by those who have seen it in its native haunts, and it is hoped that from plants and seeds somewhat copiously collected it may eventually find a much larger number of admirers in gardens devoted to this charming class of horticultural adornments." Every region he explored was viewed not alone with the botanist's searching eye, but was studied as well in its topographical and climatic aspects, as affecting its economic possibilities.

The conscious possessor of a talent for observation, he used it reverently, taking careful account of what so many would have suffered to pass unseen or fade into forgetfulness. Nor was he content to be simply receptive, but interrogated Nature continually. Often, intent on some all-absorbing quest, he would disappear from camp for a day or more at a time, still however, with the woodsman's unerring instinct, reappearing safe and sound.

Yet, curiously exemplifying the absorption of the naturalist in other than the affairs of his fellow-men, these notes contain no mention of his traveling companions, nor of any of the unique and interesting specimens of Western humanity with which he was continually coming into contact. The most warm-hearted, unassuming, and genial of men; one whose learning and humility were alike delightful, whose nature reflected the sweetness of the flowers he loved, and who was welcomed to every fireside; one of whom, as of Agassiz, it may truly be said: "where'er he met a stranger, there he left a friend." He yet made no study of man as man, caring only for hearty companionship, the warm greeting, and fervent God-speed.

Deeply affectionate, almost extravagantly fond of children, and with a sense of humor which often sparkled in his home conversation, he was yet so reticent that only the intimate few were aware of these traits in his character. With no expensive habits and almost no wants save knowledge, he looked on money as of value chiefly for the amount of this it could procure and diffuse. Devoted not only to his own species study but to Natural Science in general as a too much neglected part of the great educational field, he lost no opportunity to support its claims as against the dull abstractions of unused tongues and all exclusively textbook instruction.

Of his scientific achievements I will leave those to speak who shared in and were conversant with his labors. . . . C. R. Orcutt, editor of the *West American Scientist,* writes: "Dr. Parry discovered during his extensive explorations hundreds of new plants afterward described by Dr. Gray and Dr. Engelmann, and his name is firmly fixed in the history of West American botany. While his greatest service has been rendered to botanical science, yet horticulturists will not soon forget that it was Dr. Parry who discovered *Picea pungens,* the beautiful blue spruce of our gardens; *Pinus engelmannii, P. torreyana, P. parryana, P. aristata,* and a host of others of beauty and value. Through his zeal and enterprise many plants now familiar to American and European gardens were first cultivated. *Zizyphus parryi, Phacelia parryi, Frasera parryi, Lilium parryi, Saxifraga parryi, Dalea parryi, Primula parryi,* and many other plants of great beauty or utility bear his name in commemoration of his labors and worthily do him honor.

"No name is more intimately connected with the flora of West America than is the name of Charles Christopher Parry. For nearly fifty years his indefatigable labors and explorations in the West have enriched our botanical lore. His name is associated with many pleasant memories in the mind of every one who was so fortunate as to know him personally. Since 1882 he has published very important papers on the species of *Chorizanthe* on the Pacific Slope; on the genus *Arctostaphylos* (the manzanita); on Pacific Coast alders; and later, on the genus *Ceanothus,* which contains the numerous mountain and coast shrubs known as wild lilacs. These papers were the result of special studies in the field of these difficult groups of plants and contained descriptions of many new species."

Dr. Parry's work on earth is done. His was a busy, useful life, unselfish, but crowned with the proudest success. His name "has been stamped upon the mountain peak and traced in lines of beauty in many a mountain flower." At last, the gathering hand has been gathered; the wandering feet have brought him back to lie down on the green hillside within sight of the home he loved; to rest under fragrant, clustering flowers, where in years long past he was wont to seek their shy, wild sisters. They and he are gone; but let us trust with his friend, the prose poet of the Yosemite: "he has but gone botanizing in a better land."

Setting the Stage: The Parry Narratives 1

Parry wrote a series of popular articles in the *Chicago Evening Journal* describing the route of his travel from Iowa to the Rocky Mountains. The material elaborated in these articles may not impress a modern reader, but one must remember that this is virtually the first unvarnished description of the region, intended to acquaint midwestern readers with the qualities of the landscape and its potential for habitation and agriculture. Unlike Bayard Taylor's narrative (Taylor 1867) covering the same territory, from Denver across Berthoud Pass and down to Hot Sulphur Springs in 1866, which dwelt primarily on the rigors of the terrain, the hardships of travel, meetings with Indians, and narrow escapes during flooded stream crossings, Parry's is mute concerning either his fellow travelers or the difficulties of the trail. One would like to have more information about Parry's mode of travel, the amount of collecting equipment, and how the party was able to press and dry their plant specimens and carry such an enormous load for so many miles along primitive wagon roads.

The following transcriptions are made from faded typescripts, possibly made from clippings of these articles. The anonymous transcriber could not read some of the words; these spots in the text are indicated by brackets. When possible I have inserted what seems to be the appropriate missing information into the brackets. I am indebted to Deborah Q. Lewis, curator of the Ada Hayden herbarium at Iowa State University, for her discovery of this material and for allowing me to use it. The articles transcribed here would be very difficult to find otherwise.

Parry's early visits to Colorado took place during the Civil War. His route took him into Colorado by way of the Platte River route, which at the time must have been fairly free of conflicts with Indians, since he does not mention any such trouble. A few short years later, however, Indian wars were stimulated by the rapid settlement by whites, particularly in Kansas, which resulted in failed treaties and the eventual confinement of the Indian tribes to reservations in Oklahoma. The City of Denver was five years old when Parry first visited it. The great flood of Cherry Creek, which destroyed much of the city, took place on May 20, 1864.

The Far West: External Features and Natural Resources

Realizing how little is known by the general mass of our people regarding the real character, extent, and resources of the vast region of our domain lying between the Mississippi and the Pacific Coast, I propose, in a series of articles, to give the result of an extended journey and of careful explorations in that region, from which I have but recently returned. In order to begin at the beginning, I will devote the first number of this series of papers to Central Iowa, which, along the line of western emigration to the Rocky Mountains and the Pacific, has received but little attention from the professional tourist and, while occupying a considerable space on our maps, is passed over without awakening any definite ideas of the true natural features of the country which it is intended to represent.

The emigrant, having his eye fixed upon a more distant goal, is inclined to look upon it as an impediment in his journey, the more rapidly passed over the better; the man of business avoids it by taking circuitous but more expeditious routes, while those that occupy its numberless hills and valleys, or delve in its fruitful soil, are content to look upon it as the field of their daily drudgery, and leave the future to develop its claims to more extensive regard. Still, it is a country whose undeveloped resources must be apparent to the most casual observer, foreshadowing a grand historical future. Iowa has peculiarities of soil, climate, and location such as must forever mark the [ ] and character of [ ] and its civilization. It [ ] of country, not only capable of holding within itself an immense population but also of contributing largely to enterprise in the production of export. So thought your writer as, after a protracted journey of 350 miles westward from the Mississippi, he wound down the bluffs that bound the bottoms of the Missouri, opposite the mouth of the Platte River in Nebraska. If any of your readers wish to know the observations upon which the above conclusions are based, let him follow me in a rapid tour of the country thus passed over. Let me endeavor to condense my experience of each day's observation into as many minutes as the pages of your journal.

The western bank of the Mississippi, bounding Central Iowa, is too well known to require any detailed description. The gently sloping bottomland, based on a rocky substratum of limestone, here and there exposed in mural faces, is indented at variable distances by wooded cliffs, presenting at their exposed points highly picturesque views of the sparkling waters and verdant islands of the Upper Mississippi. At the foot of these bluffs, and creeping up their sides, nestle the embryo cities of this region, vying with each other, in beauty of site or commercial facilities for shipment, east and west. From the bluff summit stretches inland an undulating prairie region, exuberant in the fertility of its soil, well watered, and mostly under cultivation. From this point westward the same character of country continues, varied somewhat in details; undulating prairie ridges forming the divides between numberless valleys with timbered bottoms and tortuous streams winding between alluvial banks. The prevailing timber on the sides of the bluffs and upper bottoms is oak of various species, while on the bottomland is a more dense growth of elm, maple, ash, black walnut, willow, and linden.

Toward the west, the country gradually increases in elevation, the main water courses having a very uniform direction to the southeast. With the gradually increasing elevation of the country, the prairie ridges become more extensive, the stream bottoms narrower and less thickly timbered, and the general aspect of the country more open. The process of settlement becomes here well marked; the new settler plants his cabin on the edges of the timbered streams and runs his cultivated furrows in the virgin soil of the open prairies adjoining. Thus, these open spaces become gradually encroached upon on all sides, till later settlers are forced to take over the open prairie and provide their own shelter by the slower process of planting groves. More rarely, some adventurous colony, lying along the line of a projected railroad, will plant its town site on the open prairie, radiating thence in every direction, having ample room in which to expand, and often as ample time in which to realize its growing expectations. The staple productions of this region are wheat and corn, the former being principally relied on for export, while the latter, in connection with oats, root crops, and hay, are employed for home use or in the fattening of livestock. The progress of settlement during the past ten years has been exceedingly rapid and of a permanent character and, although the rapid growth has been somewhat checked by the collapse of speculation and partial failure of crops, yet every year still witnesses a permanent advance. New land is under cultivation, the disappointed speculator seeks to realize a small return for his extravagant outlay in the surer products of labor, while the farmer, burdened with debt, struggles the more manfully to relieve his homestead by increased toil and extended cultivation. All these will tell in time upon the permanent progress of this country, and it will eventually come out all the better for the fiery ordeal it has been called to pass through.

After leaving the geographical center of the State, the general features of the country become more open, the prairie swells are bolder and more massive; the soil, though still fertile, is exposed to fierce sweeping winds, stunting the vegetation. Settlement is confined almost entirely to sheltered valleys. Still, even here, thriving villages occupy the banks of clear, meandering streams which afford good water power for miles; adventurous farmers turn up the valley sods with their gleaming plowshares; the lowing of cattle and (too rarely) the bleating of sheep, adds a rural sound to this otherwise too desolate region. The dividing ridge, which separates the waters flowing east to the Mississippi and west to the Missouri, is not marked by any sensible changes either in the composition of the soil or scenery, but proceeding westward the valleys become more open, the streams are less timbered, and the general aspect of the country is more open and exposed. This western section is also more subject to protracted drouths, and the products of cultivation are consequently less reliable. Still, it is by no means a country to be avoided by settlers. In some respects it has manifest advantages over the more central portions in having a nearer market for surplus productions in the valley of the Missouri; its salubrity is unsurpassed and, lying as it does on the borders of a more desolate western region over which a restless emigration is constantly pouring its floodtide, it will be the first to reap the golden harvest of eastern mining enterprise.

The general impression which this entire region is calculated to leave upon the mind of the observing traveler is that of partially developed resources—Nature, rich in capacities, waiting for the plastic hand of art and culture to mold her productions into the varied forms of benign civilizations.

Those of your readers who may wish to accompany me farther on my western journey to the Rocky Mountains may by the aid of your columns meet me again on the banks of the Platte.

Eastern Nebraska Along the Platte River

The Platte River, the principal inland stream of Nebraska Territory, pursuing a very direct easterly course from its sources in the Rocky Mountains to its exit in the Missouri, may be regarded as a type of the country through which it flows and thus affords the means of forming a fair estimate of the true capacities of the region which it waters. It is, in fact, to Nebraska what the Nile is to Egypt—one of its most remarkable natural features and one of the controlling material influences in its prospective civilization.

What, then, are the natural features distinguishing this stream from others? Its peculiarities are all embodied in the classification of its aboriginal name, *Nebraska* (shallow water); flatness, want of depth, openness, and horizontal expansion, characterize not only the river but the country through which it flows. Viewed at any point along its course, though still more apparent as you proceed upward, the river exhibits a wide, shallow bed, at least one half of its surface occupied by exposed sand bars and spotted here and there by low, wooded islands. The stream itself is absolutely without a channel and spreads everywhere its thin waters and shifting quicksand in its progress to the more turbid waters of the Missouri.

A similar character of flatness marks the adjoining country on either hand; the bottoms are low, sparsely wooded, and usually rise in gentle slopes to the adjoining tableland. Occasional high bluffs form, in fact, the only exception to the general feature of evenness of surface. These, at different points, exhibit steep hills cut up by entering ravines. But on reaching the summit a short distance back, the same character of flatness, on a higher level, is exhibited in the undulating plains that sweep the horizon on every side. Still, as before indicated, this level character of country is not the result of a sudden change in the aspect of scenery as before noted on the eastern side of the Missouri; the transition is gradual and almost insensible, from the undulating prairies of Iowa to the level plains of Nebraska.

The first fifty miles from the Missouri westward, the same character of hills and valleys continues as before described in western Iowa, but an observing eye will soon detect less abrupt elevations of surface, a gradual flattening down of ridges and hills, a tendency to spread out into undulating plains, more scarcity of timber, and a more even horizon. The soil still maintains fertile character, but the vegetation indicates more aridity of climate, and the timber growth is stunted and inclined to form a bushy growth. Finally, *the plains* proper make their appearance, still undulating but inclined to form level stretches, and thus the characteristic Nebraskan features become plainly manifest. The increasing elevation of country as you approach the mountains serves to exaggerate the expansiveness of surface in a remarkable degree. The river seems to run almost on a level with the plains; timber becomes scarce, even on the borders of the main stream, and is mostly confined to islands. The soil grows more arid and supports only the short, tufted growth of aridity-loving plants, and for all purposes of cultivation the land may be properly denominated as sterile. Such is a fair exhibit of the natural features of the region of country under review, designated as Eastern or Central Nebraska.

In concordance with the general plan proposed in these sketches, it remains to inquire: What are the effects which these natural features are calculated to exert upon the civilized aspect of this region? Obviously, the first view to be taken of this region as a whole is its lack of adaptation to agricultural purposes. Thus, with the exception of a narrow belt having an average width of fifty miles west of the Missouri and perhaps extending somewhat farther up the valley of the Platte, in which the conditions of the soil and climate may favor a limited agricultural capacity, the entire region of Nebraska proper may be regarded as unfit for ordinary field culture. This is due, as we have had occasion to remark, more to the peculiarities of climate, which are beyond the control of human agencies, than any absence of fertile elements in the soil itself; hence, seasons may occasionally occur where a partial success may follow the labors of the pioneer husbandman, but this will only prove a rare exception to the general rule of relative sterility stamped by the hand of nature on this wide scope of country. This is clearly shown in its peculiar native vegetation, consisting mostly of stunted forms with long penetrating roots and susceptibilities of withstanding great vicissitudes of dryness, heat, and cold. The country capable of producing such forms will never yield to the more exacting requirements of the plow and harrow. Even civilized weeds, which nearly everywhere follow emigration, fail to find a foothold on these uncultivated wastes; only a few straggling specimens, scattered here and there by the wayside, serve to remind us of Eastern homes.

What branch of human industry, then, is adapted to develop the requirements of this region, and how far is it susceptible of the ameliorating influence of civilization? It must at once occur to every thoughtful mind that the only profitable channel in which labor can be exerted in this region is in the rearing of animal products. Looking over the rich, green sward which, in the early summer months clothes the undulating plains and broad, open valleys of the Platte, the first exclamation will be, what a fine pastoral region! Here the herder and the shepherd can spend a listless life in merely watching the growth and increase of his flocks and herds. This, in fact, would express the true state of the case if the vicissitudes of climate did not impose restrictions upon such uninterrupted luxuriance. Summer drouths and keen winter blasts must be first taken into account, and then the fact may be fairer stated that this is pre-eminently the pastoral region of this continent. The nutritious character of its native grasses, the facilities for watering stock, and the almost unlimited range, offer everything that could be desired for yielding the largest returns from the smallest expenditure of labor and capital. Where the buffalo has heretofore roamed in countless herds, and where the antelope abounds, more domestic races may be reared with still greater facilities. It is even yet a question unsettled, whether these wild denizens of the plain may not be educated to the service of man and thus increase their usefulness. It would well deserve the attention of a humane government to institute experiments for settling this problem. However this question may be finally answered, the capacity of this region under consideration for grazing purposes remains unquestioned, and on this fact will hinge the character of its future civilization.

That pastoral pursuits will be the predominating, if not exclusive, branches of human industry pertaining to this region, is further evident from the entire absence of mineral wealth or motive power suitable for manufacturing purposes. With the exception of the Missouri River, it has no navigable streams, the Platte through its whole course serving, as far as any useful purpose is concerned, only as a huge watering trough for men and animals. The indifferent timber growing along its banks would hardly suffice to fence properly the limited agricultural district which it occupies.

On the other hand, while the Platte River is entirely unfit for commercial purposes of transportation, the valley affords the most eligible site for the location of a great East and West railroad. The necessary grading and bridging through its whole extent would be comprised within the lowest estimates for such kind of labor; miles of track could apparently be laid without the necessity of turning over a shovelful of earth. This is a very important fact in reference to the future growth of this region, which will require importation of nearly all the necessities of life and make returns in animal products.

From the above facts, we have the data for estimating the future civilized history of this region: a sparse human population with few fixed habitations, leading a nomadic life, following here and there, according to the exigencies of the season, immense herds of cattle and flocks of sheep; lordly proprietors reckoning their wealth by the countless hooves that roam over unrestricted fields; a patriarchal style of living; an impatience of control; an attachment to the natural as opposed to the artificial; and, as we would fain hope, a patriotism planted deep and growing strong, like the deeply rooted plants of their broad, undulating plains. Such, may we not fairly estimate, will be the future population of Nebraska Territory.

Western Nebraska and Eastern Colorado to the Foot of the Rocky Mountains

It is difficult to characterize the region of country above limited by any one general feature, otherwise than as an elevated, arid region, diversified by considerable differences of surface elevation, varying somewhat in the nature of its soil, but all agreeing in the uniformity of its native vegetable products, its unfitness for ordinary cultivation, scarcity or entire absence of timber growth, and a general lack of those conditions which render it suitable for civilized habitations. It would be difficult to draw a precise line of demarcation between the districts of country thus described, and that before noted as pertaining to Central Nebraska, especially as there exists no natural boundary between them, and the passage from one to the other is only marked by insensible changes in the vegetation corresponding with alternation of soil, a more irregular outline, and increased elevation, by which the surface drainage is disposed to form more abrupt banks and deeper channels along the various water courses. There is also a very perceptible change in surface material, being composed of coarser particles, the light alluvial earth of the Nebraska plains giving place to sand and, as you approach the mountains, to coarse gravel evidently derived from the detrital material composing the mountain range. All the stream beds, whether supplied with constant water or depending on uncertain rains, are bedded with coarse sand or gravel and show less of that quicksand character so noticeable along the lower course of the Platte. All these changes are the natural result of the approach to the great western barrier of mountains which has such an important bearing both on the climate and soil of all the region watered by its numerous streams.

Most of the peculiar features of this region may be best summed up in the natural vegetation, which thus affords the best key to its resources and probable future history.

The first point that will be likely to attract attention in this respect is the appearance and prevalence of cacti. This class of plants always indicates a coarse, porous character of soil, in which the roots penetrate, apparently more for the object of maintaining a foothold than as a means of deriving nourishment. They thrive in a clear, dry atmosphere and are capable of withstanding very considerable vicissitudes of heat and cold. Though not occurring in great number or variety, owing to the severe cold of the winter season, they include one or more of all the extra-tropical genera, several of which find here their northern limits.

What the cacti are to the soil they occupy, so are the native tribes that inhabit it. The crow and the buzzard visit it only to regale on the carrion it affords from worried beasts of burden who have dropped under the yoke and left their forms, with a shriveled skin adhering, to become the dry mummies of the plain. The antelope skips over it, taking advantage of its fleetness to gather up its richest uncropped grasses. The prairie-dogs find here a congenial site for their towns, and, perched on the top of their respective mounds, salute the traveler with their sharp social bark. Having thus succeeded in attracting attention, down they hop into their holes with a coquettish flirt of their fan-shaped tails.

Even the Indians have no fixed habitation but, scattered about in small camps near some water course or tepid spring, exhaust their stock of antelope meat and then either beg the means of subsistence from passing emigrants or move to some other less frequented locality. Everything, whether animal or vegetable, is evanescent in its character and either moves or shrivels away. This must always, under present conditions, be the character of the future inhabitants of this district. The only permanent landmarks in this region will be what commerce requires in its stage stations soon to give place to the telegraph and the locomotive. These annihilators of space and time will accomplish for this country all it is capable of supplying, in connecting the mineral wealth of the Rocky Mountains with the agricultural fields of the Mississippi valley. The region through which these civilizing influences pass will gather up the scattered gleanings of these rich harvest fields, dropped by the wayside and may well be content therewith.

But the above view, though true of a large scope of the region under consideration, would not be complete till we take into account a small strip of country immediately adjoining the mountain range. There this gigantic barrier breaks the monotony of the plains that skirt along its base, like the mainland to a shore line, so does it send out reefs and islands to diversify the nearest surface of this broad expanse. This, of course, has a modifying influence on the climate, soil, and productions of this connecting belt of country. To make amends for increased elevation of the general surface, it affords shelter from the bleak, sweeping winds of the outside plains; it catches a share of the mist and rains that bathe the mountain slopes; it receives into its arid bosom the more copious tribute of melted snow from the higher elevation of the snowy range; it occupies a break in the progress of commercial transportation such as favors the growth of large cities. Here long caravans loaded with Eastern products deposit their cumbrous loads, and animals, wearied with the long, tedious march of six hundred miles, recruit their energies by cropping the nutritious forage that beds the grassy slopes or shoots up its verdant spikes in clear-watered valleys.

Here, too, cultivation offers valuable returns for properly directed labor. Doubtless, when the peculiarities of climate and soil are better understood, less uncertainty will attend upon this branch of industry. But, still, it must generally be confined within narrow limits and be mainly directed to the production of the more perishable articles of diet or such as are unsuited for distant transportation. Under the shelter of these steep ridges and grassy slopes, at the base of the mountains, cattle will find shelter from the inclemencies of winter storms, and the wandering herdsman of the plains will here find winter quarters for his roving herds.

All these various conditions will combine to modify the civilized history of this region, when the great American enterprise of a Pacific Railroad shall have been consummated, and the dweller upon the distant shores of the Pacific shall stretch a hand of greeting to the inhabitant of the Mississippi valley. Then this halfway house of the continent will be the gathering place of a mixed assemblage of various callings and pursuits, intent upon the gains of commerce, or pushing forward the great enterprises of civilization, developing natural resources, erecting schools and churches, and carrying on all the manifold machinery of social life under the shadow of that great mountain mass where the setting sun gleams upon fields of perpetual snow and where the rippling streams, flowing to either side of the continent, are freighted with the clearest and coolest beverage that nature bestows except, when disturbed by human agencies, the industrious miner loads them with the refuse of gold-bearing gulches or powders them with the crushings of quartz mills.

This brings me to the more special examination of the mountain slopes, which have been the main object of our journey, whose peculiarities will be noted in subsequent communications.

First Impressions of Rocky Mountain Scenery

It was certainly a very questionable exercise of good taste, to say the least, by which the unmeaning title of *Colorado* was applied to the territory now known under that name. A country possessing in so marked a degree such expressive and prominent natural features would seem to deserve a significant and appropriate name. Aside from the euphonious flow of letters, which might have been better supplied from some of the aboriginal dialects (Arapahoe, for instance), the present name is suggestive of nothing properly pertaining to the region to which it has been applied. *Colorado*, meaning simply *red*, is in no sense characteristic either of the country or its products. Its gigantic mountains, capped with eternal snow, convey, in their bright, reflected light, nothing corresponding to the name. The intense blue depth of its skies is equally inappropriate, while its clear mountain streams, its evergreen forests, and the most usual gray shade of the alpine rocks, answer to any other color better than that of *red!* For a worthy historical name, Jefferson would have been unobjectionable, or New Switzerland would have awakened some associated idea in this alpine region of the New World. With this protest strongly impressed on mind by a recent view of those "shining mountains" of the West, let me attempt a sketch of first impressions and subsequent experience in these picturesque retreats.

On coming in sight of the Rocky Mountains from the broad expanse of the Great Plains, their appearance is at first indistinguishable from the irregular masses of cloud that so frequently edge the horizon. There is the same mottled base surmounted by fleecy summits. To an unpracticed eye their dim outline appears at least to change according to the varied tints which they reflect, and when, as is often the case, cloud and snow combine together, the deception is perfect, and not until a nearer view detects the sharp, immovable outline of certain prominent peaks or the setting sun brings into bold relief the broken crest, can the fact be realized that the great vertebral ridge of the continent is in sight. With such an object constantly bounding the onward view and enlarging its dimensions with every hour's travel, the last one hundred miles of toilsome journey over the open plains is relieved in great measure of its monotony, and renewed vigor inspires the quickened steps of the traveler.

Near the base of the mountains, smooth, grassy swells, gigantic in comparison with the undulations met with on the Great Plains, give a foretaste of the ruder character of the scenery soon to be experienced. Projecting masses of dark-colored rock frown upon the traveler as he enters the rude defiles by which the traveled roads penetrate the mountain range. The view is soon shut in by rocky walls on either hand, and irregular broken slopes, scatteringly timbered by pine and spruce, occupy the contracted horizon. The clear, meandering stream along which the road winds crawls over slippery rocks and is mostly hid from view by the over-arching branches of alder or birch. Violets, anemones, and wood-flowers display their delicate tints in marked contrast with the arid, dull-colored forms that occupy the unsheltered tracts before passed over. Still, aside from the immediate border of streams and springy swales, aridity continues to characterize the vegetation. The shrubbery that strikes root in rock crevices is of a stunted and gnarled appearance. The thick, tufted plants send their deep tap roots far into the ground in search of the requisite moisture; and most of all, an almost entire absence of deciduous trees give to the woods a bald, wintry aspect. The pine trees are rarely of gigantic stature, yet in their symmetrical outline and tasteful grouping, present an array of picturesque scenery to which novelty adds an unwonted charm. Far from the civilized haunts, you feel like treading in a world only partly finished, and as ever and anon the rude miner's shanty or the wayside tavern comes into view, you almost expect to hear some strange dialect and foreign accent issue from the uncouth lips that greet you.

Remarkable also is the scarcity of animal life that breaks the silence of these wooded solitudes. The feathered songsters are rare, and the few occasionally met with give utterance only to faint or shrill notes, by no means musical. No creeping reptile glides noiselessly away from your approaching footsteps. The chipmunk alone seems desirous of attracting your attention by his agility as he frisks nimbly over the ground or salutes you with a familiar chirp from his pine tree perch. Silence and solitude mark these forest wilds, where even the loudest explosion (owing to the rarity of the air) awakens only a faint, short echo.

With a gradually increasing elevation, the traveler soon begins to experience some of the effects of rarefied air. This is first noticed by an unwonted shortness of breath and an involuntary tendency to prolonged inspiration and forcible expansion of the chest, thus by an increased bulk endeavoring to make up for the diminished density. In climbing up any steep ascent, a panting breath, swollen veins, and occasional bleeding of the nose require frequent and prolonged pauses such as are not laid down in the ordinary works on punctuation. To a person in health, nature soon adapts herself to this condition of a rarefied atmosphere, the lungs expand to take in a larger bulk of air, the muscles acquire vigor by exercise and thus give tone to the circulating vessels, while a steady coolness of the atmosphere and that exhilaration of the animal spirits that generally attends outdoor life makes ample amends for other deficiencies.

This fact, together with the absence of the more common enervating influences of civilized life, will account for the usual good degree of health experienced by the mountain voyager. Even to those predisposed to pulmonary complaints, where there is left a sufficiency of healthy lung to undergo the expanding process, a careful regimen and gradually increasing exercise is often productive of the most beneficial results. On the other hand, where the lungs are permanently disorganized, the rarefied air acts at once as a negative poison, and speedy removal to a denser atmosphere is essentially necessary to prolong the flickering existence of the confirmed consumptive. These facts should be steadily borne in mind by physicians in recommending to their patients "a trip to Pike's Peak for health," for in strict accordance with the nature of each particular case will the result be either beneficial or the reverse.

The habitable belt of the Rocky Mountain district extends from its base, having an average altitude of 6,000 feet above the sea, to the higher valleys and parks, reaching to the height of 10,000 feet. Within this range of 4,000 feet are comprised, at the lower elevation, a limited agricultural district mainly dependent on irrigation supplied by the mountain streams flowing from the higher elevations. These streams attain their maximum during the season of melting snow in the month of June, succeeding which is an uncertain rainy season extending through July and August. When these favorable conditions combine to furnish sufficient moisture for growing crops, garden vegetables can be raised of excellent quality and, in the larger bottomlands, in considerable quantities. But still, with our present knowledge of the variability of different seasons, more or less uncertainty must necessarily attend agricultural enterprises through the whole Rocky Mountain region.

With an increased elevation, agricultural products become more and more precarious, having to run the gauntlet between nipping frosts on the one hand and withering drought on the other. The highest points at which ordinary cultivation ceases extend to 8,500 feet elevation; all above this is only suitable for grazing purposes, while the highest permanent settlements are occupied almost exclusively as mining localities. Some of the peculiar features of this elevated region may be properly deferred to subsequent articles.

The Snowy Range of the Rocky Mountains

While civil war with its bloody hand has been spreading carnage and desolation over the mountain slopes and along the picturesque valleys of the Alleghanies, Peace and Industry have been achieving their less repulsive victories over Nature in the quiet recesses of the Rocky Mountains. To look back over the wonderful changes that have marked the progress of the more accessible portions of our country during the present century, the time appears long and advancement slow towards realizing the glowing anticipations of the sagacious Jefferson in regard to the Louisiana Purchase, made sixty years since. The early explorations of Lewis and Clark, Pike, and Long, seemed, at the time, to fail of fruitful results. But when the proper time has come in the designs of an all-wise Providence, the rock is smitten and streams of gold issue forth to astonish the nation. Now we see results spring forth as if by magic. Towns and cities occupy the former solitudes. The parched earth, moistened by artificial irrigation, is made to yield the products of cultivation. Hamlets are perched here and everywhere along the mountain slopes, and deep, yawning chasms reveal the tireless industry of strong hearts and stalwart arms. Let us, however, for the present turn aside from this busy hum to note the more quiet aspects of nature in less frequented haunts.

Clear Creek

Clear Creek, whose pellucid waters secured from early explorers its present strangely inappropriate name, is now probably the most turbid and muddy of streams, so that along certain portions of its course it is absolutely unfit for drinking purposes, but its valley, especially toward its headwaters, includes some of the most characteristic and attractive features of Rocky Mountain scenery. It pursues a very direct easterly course, deriving its waters from some of the most elevated portions of the Snowy Range. At different points along its course settlements have pushed their way to the base of the higher mountains. Most of the small towns of this district are located on expanded portions of the valley which, when of small dimensions, are popularly known as "bars" and, when of greater width, have received the more poetic name of "parks." Scattered groups of graceful pine trees set off agreeably the rough, rocky slopes of the adjoining mountains, and here and there clumps of spruce edge the course of the turbulent stream. The more open parks form even, grassy slopes, bedded with a pale green sward. The soil is composed of gravel, more or less coarse, resting on a substratum of water-worn pebbles and rounded boulders, extending at a variable depth to the bedrock. The adjoining mountain slopes are in most cases abrupt, rugged, and precipitous, either clothed or supporting in their crevices the characteristic shrubbery of this region, while the smooth, rounded, grassy slopes themselves, and high tablelands indicate an ancient water level.

The ravines cut by watercourses tributary to the main valley partake in a measure of the same general features on a smaller scale. The steeper and more abrupt ravines coming down from a limited portion of the adjoining mountain slopes, being dry the greatest part of the year, are what are popularly known as gulches. By these routes roads are constructed to the higher levels and divides, and along their courses gold mining in its various branches is carried on. Towards the headwaters, approaching the foot of the Snowy Range, numerous streams pour down their tribute of melted snow by a series of foaming cataracts, and the main stream finally divides into various branches to penetrate the recesses of the dividing ridge.

Following up one of these streams by a series of abrupt climbs, a gradual ascent is made to the bold alpine exposures which characterize the Snowy Range proper. The view at first is entirely shut in by a more or less dense growth of pines; only now and then, from a projecting point, you catch a near view of the mountain slope with its patches of snow, or overlook the deep valley from which you have made the ascent. The brook along which your course leads dashes impetuously downward, presenting a sheet of foam rivaling in whiteness the snows in which it has its sources. Along its borders and bathed in its icy waters grow a variety of charming plants not found elsewhere. The adjoining pine woods are mostly bare of undergrowth, but give place occasionally to dense aspen thickets.

All at once you come upon a bare, rocky exposure and patches of snow. The valley opens, trees disappear altogether or are reduced to prostrate forms, and the bold alpine exposures rise majestically before you. With this change of scenery, new plants of the most brilliant colors are seen to cover the alpine sward, flowers bloom in close proximity to banks of snow, and rock crevices, adorned with some peculiar and unique form of vegetation. You seem introduced at once into a new world. The mid[ ] that you left below is converted into [ ]ery springs, the icy brooks are edged with snow or occasionally disappear under snow arches beneath which they are heard to gurgle along their unseen course. As you climb the alpine exposures the distant scenery opens to view in distinct outlines. From some commanding height on the crest of the Snowy Range, the eye takes in a grand profile of distant snow mountains, deep, sheltered valleys, and the head sources of streams flowing respectively east and west, having their origin in encircling banks of snow or emerald lakes girdled with midsummer ice. Still, even with all these wintry aspects, there is a sense of summer in the warm rays of the sun that strike fairly within some sheltered nook. Flowers continue even to the highest crests, exhibiting their peculiar vivid colors. Insects buzz about you, and the Siberian squirrel makes his salutation from the safe shelter of a pile of rocks.

But anon your summer reverie is disturbed by a gathering cloud that sweeps over the surface and sends its chilly blasts through your system, warmed by the exertion of climbing, sputters of snow or pelting hail following in its wake. Occasionally wide stretches of alpine meadow furnish a smooth, grassy sward, doubly grateful to the foot rendered sore by the rough, rocky ascent by which you have attained this commanding position.

Here, at an elevation of 12,000 to 13,000 feet above the level of the sea, the mountain sheep and goats [mountain goats were not introduced into Colorado until the late twentieth century] find summer pasturage, and the noble elk raises his branched antlers, startled at the unaccustomed sight of human forms. At some other point you start a covey of the Rocky Mountain ptarmigan with its spotted plumage, varying in color from a pure snowy white in winter to a motley granite in summer. All these furnish scenes of never-ending interest, till the declining sun and chilly evening air warn you to take shelter for the night under cover of the nearest pine grove. Here the withered trunks of the blasted pines supply the alpine camp with comparative comfort through the chilly night watches.

A very important part does this Snowy Range supply in the general economy of a vast range of adjoining country, while its highest peaks, reaching an elevation of over 14,000 feet above the sea, do not in any case attain to the zone of perpetual congelation. The accumulated snows of winter, heaped in massive drifts, form the sources of streams that supply verdure to the low, parched valleys and open plains exposed to summer heat. The same cause operates to condense the moisture carried by ascending currents of warm air up the mountain slopes, depositing it again in the form of summer rains or autumn mists, to continue the fertilizing circuit. Even on these frozen wastes where a rigid winter holds its sway for months continuing, the accumulation steadily increasing through this season gives protection to the delicate plants that luxuriate in these summer days, while these again supply nourishment and warmth to burrowing animals, and winter pasturage under this canopy of snow. The larger animals, such as the elk, the goat, and the mountain sheep, return to the lower valleys in winter, to return to this choice summer pasturage with the retreating snows.

Here, then, far above the accustomed haunts of men, we find a charming pastoral region which will yet be made to yield rich dairy products surpassing, it may be, the world-renowned cheeses of the Alps, and delicate fibrous tissues whose fabrics may yet rival those of the looms of Cashmere.

The Route of the Pacific Railroad

It would be obviously impossible, within the short compass of a newspaper article, to do justice to a subject involving so many and such varied points of interest as are connected with the progressive construction of the great national work of a Pacific Railroad. Still, it has such a direct and immediate bearing upon the development of the resources of the Far West and is destined to form such an important link in its future history, that it would imply still greater injustice to omit a notice, however brief, of its present and prospective influence on the region of country through which it is to be located.

I must, however, content myself with alluding only to a few of the more general and less disputed facts connected with this great work being such as would naturally occur to a thoughtful observer who has studied to some extent the peculiarities of the region, and, still better, such as are embodied in the settled opinions of intelligent scientists and explorers who have felt the effort well weighed the results that are not only desirable but attainable by such.

The first point that may safely be laid reference to the disputed question and that may thus be acted: That the most direct, practicable line should be followed passing through the established centers of population and traversing, as far as may be, a habitable and desirable region for future settlement. This statement, I believe, embodies the science of all successful railroad enterprises, whether great or small, insuring at the same time all the advantages that result from the nearest connection of extreme points and the very essential fact of securing the largest amount of local way traffic. Practical difficulties in such a plan will [require us] to reconcile any conflicting views as to how far the main line should be deflected to overcome, at least, expense, engineering obstacles, what should be considered main points on the route, and where near the direct line the most habitable region is to be found.

In regard to the first point, engineering obstacles, we have seen that on the first portion of the line, extending from the Missouri to the base

C. C. Parry (undated). *Courtesy of Hunt Institute for Botanical Documentation, Carnegie Mellon University, Pittsburgh, PA.*

of the Rocky Mountains, there are no formidable obstructions and, as far as grade is concerned, the line can be located anywhere, with an average ascent of seven feet to the mile. I have indicated a preference to the Platte Valley because it is the largest stream in a direct line and its valley has been already selected as the natural route, marked out in the progress of emigration and settlement. Furthermore, in a region exclusively devoted to pastoral pursuits, there will be no occasion for deflecting the line, which can be accordingly located with exclusive reference to engineering requirements.

But to what established business center shall this line first be directed, to comply with the conditions laid down in our general statement? I answer without any hesitation—Denver City. This is undoubtedly the settled metropolis of the region lying at the base of the Rocky Mountains, as it is the natural concentrating point for the mining enterprise of Colorado Territory. Furthermore, this location is steadily assuming an important position as a diverging center for trade and travel to the south and north. Here terminate in its western direction easy grades and straight lines; thenceforth engineering skill will be required to accomplish steep ascents and short curves, involving a change in the appliance of locomotive power. Doubtless, with an increased development of the agricultural resources of the belt of irrigable land at the base of the mountains, branch railroads will sweep around on either hand to gather in these valuable products to a central depot, from which again supplies will be drawn for distant points to meet the urgent wants of less accessible settlements.

According to the laws of trade, such a central point for business must be established at the base of the Rocky Mountains, and all indications thus far point out Denver City as such a point and, consequently, one of the main stations on the great continental highway. That this is the settled opinion of the most intelligent businessmen is sufficiently indicated by the large investments of capital and expensive improvements looking to a progressive, permanent growth. Not to dwell on this point, however, which seems to require special notice at this time only in view of a general impression that a more practical engineering route may be found lying considerably north of Denver (which, however, has not been determined by actual survey and in regard to which grave doubts exist in the minds of those best informed). There is no practical doubt that a direct line west of Denver, to Salt Lake, passes through a richer mining region and an incomparably better habitable country than any other diverging either north or south. Granting, then, that the engineering difficulties on this direct line are great, the grandeur of the undertaking and the important results to be attained warrant the largest expenditure of skill and means.

It is, indeed, a fact worth considering, whether looking exclusively for easy grades and running out buffalo trails, surveys should now be mainly directed to determining the most abrupt and narrowest portions of the mountain mass through which a tunnel may be excavated, thus solving at once a host of minor difficulties and carrying out boldly the grand programme of a direct line. Recent mining developments give a special interest to this point, to say nothing of the extremely valuable scientific results which would be brought to light in the progress of such a gigantic work. Thus far, gold- and silver-bearing lodes increase in richness in proportion to the depth of the mining excavation, and it has passed into a proverb among Colorado miners: "To pass through the cap is to fall into the pocket." Enthusiasts speak of finding metals in mass at some fabulous depth in the bowels of the earth but, without laying much stress on such unreliable expectations, it is still a very important practical question to determine how far the rule holds good that this class of metallic deposits increases in richness in proportion to the distance from the surface.

Now, if it can be shown that a practicable tunnel route may be found lying in a very direct westerly course from Denver City, passing through a rich mining district and debouching in the desirable habitable region of Middle Park, thus leading in the most direct line to the Salt Lake settlements, have we not comprised all the conditions laid down in our main statement, with the additional fact of a probable development of mineral wealth that may surely pass all expectations and give a wonderful impetus to the development of every branch of industry?

The present want in connection with the Pacific Railroad is to carry on, as rapidly as possible, the actual construction of the work over the line where no natural obstacles exist, to reach the first main station at the foot of the Rocky Mountains. In the mountains, thorough surveys should be carried on to determine the best direct route across or through the mountain range. Examinations should be carried out at the same time in regard to the country traversed, as to its natural resources, its vegetable products, its agricultural capacities, its climate and its general habitable qualities, also as to the facility of procuring necessary fuel and material for construction and repairs. Settlement should be encouraged at every available point on the route, by the preliminary location of wagon roads, telegraph stations, and manufacturing establishments of different kinds. Protection should be afforded to the pioneer settlers against the incursions of roving Indians or lawless whites. Then let the work go on, steadily, unremittingly, till the great end is accomplished that shall connect indissolubly the Far West with the Mississippi Valley, the Lakes, and the Atlantic.

Of the momentous consequences that will necessarily follow the consummation of this work, we can only briefly allude to in conclusion. The opening of an entirely new avenue of trade between remote points equally interested in an exchange of products will speedily produce a commensurate effect in building up new commercial interests, not alone at the extremes but also all the principal intermediate stations. Towards all these, converging branches which naturally extend as the country on either side becomes settled and its resources developed, will pay tribute, and thus aid in swelling the tide of traffic along the main line. The stimulus given to production of every kind will be in direct ratio to the facilities for reaching a market, and ingenuity and enterprise will put in exercise to increase and multiply the fruits of remunerative toil. Especially will this be true in reference to the undeveloped mining enterprises of that wonderful mountainous region whose actual wealth may yet exceed the dream of the most sanguine enthusiast.

But apart from all these strictly commercial results, what a wonderful diversion will be given to the vagaries of pleasure travel when these mountain solitudes will re-echo the shrill scream of the locomotive. Then the pine-clad heights, these dashing streams and sparkling springs will attract the ease-loving votaries of fashion. The toil-worn and dyspeptic denizens of metropolitan cities will seek invigoration in the clear, cool atmosphere of these alpine heights and wash away the impurities of disease or travel in these thermal springs. Here the adventurous huntsman will bag the choicest game in unfrequented haunts or track with needful caution the formidable grizzly to his lair. The enthusiastic angler will throw his deceiving bait by the edges of sparkling brooks and draw from thence the speckled trout so valuable in epicurean eyes. Here, too, the earnest votary of science will penetrate every nook where the secrets of nature may be unveiled and measure with line and plummet the unexplored chasms of the Colorado.

Still more, may we not hope that a region so long held in reserve by the hand of Providence and retained in its original wildness through the centuries in which modern civilization has had its birth and wonderful progress will, in the rapid development of its material resources, play a still more important part in giving new energy to the advancement of moral ideas and, in extending the domain of civilization and Christianity, add a crowning luster to American history?

The Far West: Its Present Wants and Future Prospects

At the risk of some repetition, I cannot close the present series of articles without presenting a general review of the present wants and future prospects of that undefined region designated as the Far West. Such a review, concisely and truthfully expressed, in the present juncture of national affairs when great historical movements and important social changes are in progress, is at least worth attempting, even should the effort fall far short of the actual reality.

The great comprehensive want of this region, as of all new districts of country, is accurate and reliable information of its peculiar features and true natural resources. Such information, derived from whatever source or however limited and partial is, as far as it goes, a positive advance made towards realizing the designs of Providence and pushing forward the onward march of history. Furthermore, whatever tends to the removal of erroneous impressions or correcting the mistakes that have gained prevalence, either from the views of partial imperfect science or the exaggeration of wild enthusiasts, proves to the same extent a remover of obstacles in the path of progress.

The first special object of industry in regard to which accurate information is desired has reference to the geography of the region as a whole, and also the geographical arrangement and distribution of its several parts. Notwithstanding the large addition to our knowledge in this respect, accumulated during the present century from the labors of zealous explorers and also as the fruits of pioneer settlement, much yet remains to be learned and many glaring errors to be corrected before a reliable and accurate map of this region can be made, such as may prove a safe guide in directing the course of travel or that will warrant the deduction of general conclusions.

What we have heretofore gained has been little more than a rapid reconnaissance, connecting with more or less accuracy certain astronomical stations in which mountain ranges, river courses, and uplands have been thrown in by guess or left to the discretion of irresponsible mapmakers. What we now want is to replace these imperfect maps by accurate surveys. We want to know the position, altitude, and extent of the intricate ranges of mountains, and thus endeavor to gain an insight into some system on which they are formed. We require a knowledge of the course and meandering of streams, their average volume of water, rate of descent, extent and character of valleys and uplands, also the distribution and extent of timber growth, the zones of vegetation and, in fine, all those items of information that can be addressed to the eye on a well-constructed physical map. Connected with this and essentially necessary to a proper knowledge of the nature and extent of its mineral resources, we need to determine the character and boundaries of the different geological formations and also to bring together those accumulated facts in Natural History that throw light on the capabilities of the region for agricultural production and for determining its habitable qualities.

When these accumulated facts shall be brought together, which can only be done in the actual progress of settlement and pioneer experience, it will doubtless be seen that a Providential hand has directed the physical development of this region which, in the proper time and in the appointed way shall contribute its share in enlarging the sphere of national activity and aid in giving variety in unity to the national character.

Among the many agencies in effecting these important results and modifying them to the most desirable ends, I would venture to give a prominent place to the local pioneer press. This powerful civilizing instrument, which not only reflects the actual state of society where it exists, but also molds it to a great extent whether for good or evil, also indicates the steady march of improvement with which it needs must keep pace or, more properly be somewhat in advance. As a medium for information it embraces the never changing, versatile features of the active present like a widespread landscape, with its lights and shades, heights and depressions, from which the observing eye can best detect the general features and grasp the entire scope of the region under review. All honor, then, to those enterprising pioneers of the typographical art who have carried into the wild recesses of the Far West the civilizing light of letters and, amid all the privations and inconveniences of frontier life have aided to mold the rough face of society and to promote the advancement of material and moral improvement.

With an accurate knowledge of the resources and capabilities of this region there will be no lack of suitable means to aid in its development. Not more certain than the law of gravitation is the process by which capital finds the place for profitable investment and, following in its train, brings with it all the appliances of commercial facilities for transportation and the application of the latent and active powers of nature to work with unfaltering energy in the service of man. Thus it will be seen that all the present and prospective wants of this region are comprised in the two items of accurate information of natural resources and actual settlement to bring them out. Whatever means and measures conspire to either or both of these results will so far aid in rendering the region to which they pertain all that is designed to be in the scheme of an All-wise Providence.

Nevertheless, there is a process according to which each particular region attains to its full development, which short-sighted human philosophy can only dimly discern in its unfolding steps and which becomes clear only as we retrace the thread of history. Thus when the early and adventurous Spanish explorers, in their search for gold and the fountain of youth, penetrated to the Far West and, with most heroic endurance explored the western wilds, visiting localities that now after a lapse of over three centuries are but just opening up to the view of the present generation—treading under the hooves of their gallant charges the dust of gold that is now just pouring its tide of wealth into the avenues of modern trade—still with all the knowledge of the country thus obtained, it was not reserved for these adventurers to develop the region from which they retired with disgust or left their bones to bleach on its arid wastes. Nor again, centuries later, when the no less adventurous trappers scoured the country in every direction in search of furs, threading all its intricate streams, and clambering along its rough mountain slopes till they knew the country by heart. Neither were these the agents by which this region was to be developed to its full capacity for the use of man and to perform its appointed part in civilized history.

Before this great design could be accomplished through long, tedious years of pioneer settlement, must the rich agricultural lands of the central basin be developed. Step by step, passing through the severe ordeal of Indian warfare, combating with malarious diseases, reclaiming by steady toil the wealth so long stored up in the decaying vegetation of centuries, the slow products of timber growth or the hidden wealth of buried mineral treasures—when these have at length reached a point where their true destiny becomes apparent and an increasing population is rapidly filling up the various avenues of steady industry, then commences the new exodus by which an energetic people, having attained to a true national character, shall overflow the unoccupied public domain, reclaim the pastoral deserts to new uses, crowd back the worthless occupation of broad plains and beautiful valleys, drag from the bosom of the earth her long-buried treasures, put running streams and wasting natural powers to useful work and thus complete the development of a vigorous national American character.

The Dividing Line Between the Agricultural and Pastoral Regions

After an interval of three years, passing once more through central Iowa from east to west, I had an opportunity to review the impressions previously made and test to some extent their accuracy by direct reference to the progressive changes brought about in that brief space of time. Slowly but steadily the railroad, that iron hand of modern civilization, was laying a broader grasp upon the undulating prairies, while still pointing westward to less inviting regions. It was interesting to notice the direct effect produced by the magic touch of these iron figures. From its extremities, at every pause, seemed to drop off little irregular towns, very much as Agassiz would describe the tumbling of boulders from the point of a retreating glacier. Then, as another advance was made, these were pushed aside to form what might very properly be termed the *lateral moraine* of settlement. In every case, however, unwonted thrift and increased regard for the ordinary comforts of life seemed to characterize the immediate contact with this civilizing agent. The houses wore a neater aspect, the grounds were stocked with choicer fruits and shrubbery, the outhouses, barns, and fences were in better repair, and culture of every kind showed evident marks of progressive advancement. Even places remote but still in the direct line of an advancing railroad seemed to experience the vivifying influence of a contact still unattained, thus by unwearied energy inviting a more speedy [growth?].

What the railroad succeeds to accomplish in stimulating and improving agricultural interests must needs be modified as it approaches a region whence any culture becomes secondary to pastoral pursuits. Western Iowa exhibits preeminently this mixed character of blended agriculture and pastoral resources. Its sparsely timbered valleys possess a fertile soil that well repays the labors of the pioneer husbandman who has been fortunate enough to secure sufficient fencing material for the protection of his crops, but the wide stretching upland divides, even in the rich luxuriance of their summer garb of wild, nutritious grasses, seem only a wasteful expenditure of vegetation, involving constant and vexatious toil to the farmer in gathering together his scattered herds.

Here, then, these contrasted interests of agricultural and pastoral pursuits call urgently for a solution of the fencing problem. Must the whole labor and constant expense required for the protection of crops devolve upon the agricultural, or cannot it be somewhat fairly divided between these pastoral pursuits? Undoubtedly, for the full development of the pastoral resources for this mixed region, a system of herding must be adopted. Men and animals must be trained for the vocation of directing the movements of restless animal life, intent on securing the means of growth and nutrition, gathering up from the earth's broad and prolific bosom the sustenance which is thus stored away in the form of animal products for the supply of human wants. Fencing may, by such a process, be limited to the protection of choice garden spots and fruit orchards, while strong and durable enclosures, like the Mexican *corrals,* may be conveniently scattered for purposes of separating different herds or for selecting such animals as may be required for immediate use.

When the railroad penetrates this section, its freighting business will need to be adapted to this branch of pastoral pursuits. Live freight and animal products will then be the raw material requiring transportation to eastern markets, to be returned in the shape of manufactured articles and such common luxuries as are best developed in more luxuriant and restricted fields where the ingenuity of crowded communities is expended in multiplying the resources for the supply of educated wants.

Western Iowa thus occupies an intermediate position and lies on the dividing line between the agricultural and pastoral regions of this continent. It thus partakes of the mixed character pertaining to both. It differs from the more western plains in being more undulating, possessing fertile valleys and more numerous watercourses. Its grasses are more rank and subsequently less nutritious than the concentrated juices developed in more exposed situations. It affords more facilities for shelters from winter blasts and late spring storms. It is, on the other hand, less adapted to the ordinary means of road transportation and will require much larger expenditures for necessary grading in railroad construction. It is a fair grain-producing region, although more subject to drought and less luxuriant in yield than the Mississippi Valley.

What it would seem especially fitted for, in the general economy of the whole country, is for the fall and winter feeding of stock required for the consumption of eastern cities. The vast migratory herds that have cropped the summer verdure of the great western *steppe*, converting it into tender, juicy meats, will need to have their muscular structure hardened and their fatty tissue concentrated into suitable products for winter use. Flocks of sheep will need shelter and food while developing their fibrous coats and providing for natural increase. Dairy products can be extended late into the autumn season when increasing coolness of the atmosphere favors their preservation for winter use. All these various occupations will give employment to the different classes of industry here naturally congregating.

The populous towns of this region will be determined by the convergence of railroads from the east and west, and the true metropolis, in spite of all rival interest, will necessarily occupy that point in the Missouri Valley that is most central to the main tide of eastern and western travel. Here, as an intelligent western writer remarks, will be the "great Continental meat stall" where will be accumulated for shipment the great variety of animal products derived from the broad extent of the pastoral region. The necessities of fuel for warming and manufacturing purposes will speedily develop the coal fields of southern Iowa, northern Missouri, and eastern Kansas in which railroad interests will again be called to plan an important part in transporting this valuable material to destitute districts. The relation which this region will assume to eastern cities and the Atlantic seaboard will be one of mutual benefits, in an exchange of desirable products by which prices of necessary commodities will be regulated and kept within proper limits. The connections subsisting between this mixed pastoral-agricultural district and the rich mining region of the Rocky Mountains and Far West will be both intimate and advantageous.

Gold and silver, the great regulators of currency, will flow eastward, passing into the hands of capitalists to be employed in still further developing mineral wealth. The tide of travel eastward will be to a certain extent delayed where supplies of goods can be obtained in the shortest time and by the most ready means of conveyance. Thus will undoubtedly be unfolded, at least to future generations, the great designs of an All-wise Providence, in so diversifying the features and products of our wide-extended common country as to bind the whole in one harmonious compact in which no part is absolutely useless but each contributes its share to the general prosperity. [Here the beginning of a sentence dealing with slavery appears to be missing], thus developing to an untold extent the comprehensive plans of the great founders of this eternal infamy, the designs of those in our day who, for the base purpose of perpetuating an institution at variance with every instinct of humanity and justice have raised parricidal hands to destroy the work of our patriotic fathers, vainly attempting thus to mar the designs of an immutable Providence.

Denver City

Nothing strikes with more astonishment the traveler, passing for the first time over the great western plains, than the wonderful progress of the principal city that has grown up during the past five years under the shadow of the great continental range. Heretofore, in leaving the older settlements in the Missouri Valley, he passed only the rough abodes of frontier life, gradually deteriorating in quality as the timbered region is left behind till only mud walls and dirt roofs remain to mark the abodes of men hardly worthy the name of civilized. But what a change is experienced when, as the rising mountains cast a deeper shadow over the narrowing interval of undulating plains, Denver City breaks on the view with its mottled array of houses, dazzling white or dingy brown, with its church steeples, its public buildings and massive brick warehouses scattered along the open valley of the Platte. Entering its principal streets, the din of business and the clang of busy labor greet the eye on every side, while neatly dressed men and elegantly attired ladies are apt to remind one forcibly of the dingy apparel and features begrimed with the dust of travel with which he makes his first entree into this metropolis of the Far West.

Through the almost necessary ordeal of fire and, more recently, of flood, this young and vigorous offshoot of modern American enterprise is still destined to maintain its upward growth and keep pace with the development of the natural resources of the region of which it is the actual business center. In the days of early rivalry between contending city sites, it was thought by many that a location nearer the base of the mountains would be the natural point of the principal city. Hence, quite a number of rival towns located directly at the foot of the mountains claimed, at different times, the distinction of *metropolitan.* Most of these have dwindled down to small villages or rural farms while Denver, in spite of rivalry and reverses, has still maintained its onward progress. Its location in the valley of the Platte, nearly fifteen miles in a direct line from the base of the mountains, enables it to stretch radiating arms toward different points of the mountain region by which it naturally concentrates travel and forms the central base of supplies for a wide scope of mining country. Doubtless other localities, equally advantageous and perhaps more desirable in many respects, might have been selected that would have answered the conditions of a business center. But, having now attracted permanent capital, it is hardly probable, even in spite of late reverses and to some extent an unwise, illiberal policy, that it will ever be supplanted by a more successful rival.

It is interesting to notice in this connection how the principal town naturally reflects the character of the entire region to which it pertains. Mining interests are of all others the most uncertain, most subject to sudden and unexpected fluctuations. Hence, a mining population is the most excitable, the most easily led away with wild rumors, sanguine beyond all reasonable expectation or, on the other hand, unwarrantably depressed. This state of things naturally infuses itself into the character of the people and is evidenced every day in the streets of Denver. All trading is of the nature of speculation. Prices of all commodities fluctuate by the hour. A mere rumor will suddenly raise or depress the price of flour one-half. You may see a person in the streets in the morning anxious to sell his horse for $100, and the same person will refuse twice that sum in the afternoon. A mining claim may be offered today for $10,000, and tomorrow the broker may receive orders not to sell at less than $500,000. Steady, productive labor, and reliance on small, secure profits, form exceptions to the general rule in which eager speculation is the order of the day.

There are manifest advantages and disadvantages connected with this state of things. In the violent commotions thereby occasioned, sudden fortunes are made and freely spent in the way of improvements. Individuals disappear from sight but their works remain a permanent addition to the wealth of the country. Capital is invited where there is a prospect of rapid increase. On the other hand, the despondency of losses, the natural aversion to steady, productive labor, and a passion for gambling in all its various forms, is encouraged, the natural results of which are manifestly injurious to the best interests of the whole community.

There are several needed and easily obtained improvements that would add much to the attractiveness and health of Denver. First of these is a copious supply of pure water. At a comparatively small cost, an abundance of this necessary article could be furnished to every house. Numerous fountains and running streams might diffuse coolness in the heat of the day or assist in furnishing necessary drainage. The present dry wastes surrounding the town, and the bare, desolate enclosures that give such a forbidding aspect to the dwelling houses, might be converted into gardens and flower beds. The branching pine might be made to spread its delicious shade along the public avenues, and the spruce of the mountains to raise its graceful spire by the side of the church and the school house. All that is needed is a sufficient experience of the nature of the climate and the capacity of the soil to render Denver City as attractive in its internal arrangements as it is grand and sublime in its surroundings.

The future growth and progress of Denver will naturally depend upon the development of the resources of the country to which it stands in the relation of a basis of supply. Whatever tends to aid the latter will of necessity push forward the former. In this respect it has unquestionably the fairest prospect of future greatness. The untold wealth of the mountains, if made to pass through her avenues, must leave in its train all the elements of successful commercial prosperity. If, further, it should be brought in the direct line of the great Continental Railroad it will be one of the points the earliest to realize the beneficial results of the great enterprise. Make it accessible by easy means of travel from the Mississippi Valley and the remote East, and here will be a grand central point for the interchange of various products. The comforts of civilized life can then be realized here as well as in older regions. Whatever is especially desirable in its peculiar climate or conducive to health will attract travel and permanent settlement. Every mountain road thence constructed, penetrating the fastnesses of the mountains, will yield its tribute to the general prosperity. Every bridge made passable to travel will pay toll at the center of business, the larger in proportion as it is made the more free. Hence, the true interest of such a city claiming to be metropolitan is to encourage, by every possible means, travel back and forth. And the miserable policy lately in vogue, of extorting from the necessities of travelers exorbitant charges or enterprising unnecessary delays is not simply unwise but ruinous.

Still one other point which would add materially to the attractiveness of Denver City would be the establishment and maintenance of an institution of learning, not only where the youth of the country might be educated to fill important and useful positions in life, but where libraries and museums of natural history may accumulate material for studying and illustrating the natural production and resources of this most interesting mountain region, where meteorological observations can be kept, exhibiting the true climatic conditions of this region, and affording a means of comparison with less favored districts. Such an institution would prove attractive to men of science all over the world, and secure, by means of exchanges, valuable material for purposes of instruction and public interest.

Looking at Denver City as it is, however, with much that is wanting and not a little to condemn, there is still more to encourage. With a free press and a sanctified pulpit, evils will be in due time corrected, and the great fact of establishment of a civilized American city in this Far West region, supplanting the domains of barbarism and giving much promise of future greatness, is not simply a matter of present congratulation but more of future hope.

Empire City

December, 1864. In the heart of the Rocky Mountains in Colorado Territory, a small collection of unpainted wooden houses surrounding a liberty pole and a public well, has been dignified by the high-sounding title of *Empire City*. Whatever is lacking in imperial splendor in the *artificial* accompaniments of this location is more than made up by the majestic natural surroundings that meet the eye on every hand. Rugged pine-clad ridges rise abruptly from the open valley and stretch in irregular lines, increasing in bulk and elevation, till they terminate in the bold, bleak summits of the Snowy Range. Through this irregular and broken section of country the natural drainage is distributed in a great variety of forms, including the dashing cascades of the mountain slopes and the rapid winding streams of the main valleys.

The town itself occupies an irregular grassy slope on the northern banks of Clear Creek, thence struggling up a small branch valley known as Lion Gulch. It terminates in the elevated mining location called Silver Mountain. Here the various branches of mining industry have made their mark on the natural face of the country, denuding the soft surface materials and exposing to the light of day its long-buried mineral treasures. Looking south from this elevated spot the eye is attracted by the fresh verdure of Bard Creek Valley, resting in quiet beauty in the lap of rugged mountains, hemmed in on the one side by the rough, lank form of Mount Lincoln, while directly opposite, rising in smooth, grassy slopes and crowned by noble evergreens, is Mount Douglas. Following up Clear Creek, which here flows nearly direct from the west, within whose deepening gorge the setting sun sends forth his parting rays, we first pass over level stretches of upland overlooking a wide bottom occupied with willow thickets and a maze of deserted beaver dams. An artificial dam thrown across the main stream for milling purposes has collected the waters into a clear lake, prettily located at the outskirts of the town, while just below this Clear Creek loses the appropriateness of its name in receiving the turbid waters flowing down Lion Gulch from the upper mining location.

In the town proper, the Snowy Range is hid from view by intervening mountains. Bald Mountain, with its patches of snow, comprises the most conspicuous alpine aspects of this location. From Bard Creek, however, a fine view is obtained of the bare slopes of Mount Flora and the snowy crest of Parry's Peak. The latter, situated in the dividing ridge, has an elevation of 13,000 feet above the level of the sea and commands an extensive view of the Great Plains on the east and Middle Park to the west. About a mile above the town we come upon a sawmill, conveniently and picturesquely located on the lower side of a rocky slope, deriving its water power from a dashing mountain stream appropriately named Mad Creek. The adjoining valleys and mountain slopes furnish the necessary pine timber for conversion into the different varieties of lumber, mostly of small size but of excellent quality.

At this point the last settlements are met with, and over the wide scope of country beyond to the west, Nature rules supreme. There the works of man disappear from sight, natural attractions come more distinctly into view. Leaving the beaten track, by which during the later summer months travel is directed towards the different points of interest across the range, you pass at once into the profound solitudes of unfrequented woods. Here you encounter a succession of rocky slopes, adorned in their season with a variety of charming plants, now blushing with the luscious richness of the creeping strawberry or later displaying the profuse cluster of the mountain raspberry. The steep, rocky ridges lead to high, level benches of ground from which, through the open timber you catch occasional glimpses of the deep valley beneath, with its meandering watercourses or the bold alpine summits above, flecked with patches of snow. Another rather severe climb and the limit of tree growth is reached, where the varied attractions of alpine landscape are brought distinctly to view.

These several features, all comprised within the compass of an easy day's travel, serve to render this remote settlement a point of special attraction to the admirers of nature. Hence, in the late summer months unwonted life and activity are exhibited in these otherwise quiet streets. Men of business seek invigoration in the cool atmosphere from the oppressive heat of the lower valleys. Hunting and fishing parties here make up their outfits for the rough journey across the range to Middle Park. Later still come the fruit gatherers, intent on securing their winter stores of these desirable luxuries, till winter, coming down with severity from the alpine heights, hurries the belated traveler back to winter quarters.

The town proper has an elevation of 8,800 feet above the sea, being thus 2,000 feet or more above the highest summit of the Alleghanies. This gives an atmospheric pressure equivalent to 22 inches of mercury in place of 29 inches, which latter is the usual average of pressure experienced in the most thickly settled portions of the States bordering the upper Mississippi Valley. This is not very sensibly felt at first by most persons in sound health, but a prolonged residence is apt to beget a degree of lassitude and inaction quite noticeable in old settlers. Horses and cattle also experience the same debilitating effects, so that less than the usual amount of work is accomplished in a given time, whether by men or animals. The common laboring man goes to his day's work with a sort of *drag*, and whatever is required to be done is gone about leisurely. Hence it is that a slow and easy style of life is prevalent. The miner and the artisan finish their day's tasks and then spend their leisure time in "lying about loose" or engaging in light, active amusements. Such, I suppose, will be more or less the character of the inhabitants that may choose this elevated altitude for a permanent home.

The social history of the place is brief and possesses little of startling interest. The first cabin erected in this vicinity was located at the foot of Douglas Mountain late in the year 1859 by a pioneer settler named Bard, who located a *ranch*, or farm claim on the lower part of the valley that now very appropriately bears his name. Becoming soon dissatisfied with his isolated position, he sold out to a later settler who has till recently occupied the place, deriving considerable profit from it as a grazing and dairy farm. In the year following (1860) a party of prospectors discovered the rich mining lodes of Silver Mountain, and a town was started near the present location and called Valley City. Speculation in town lots not proving immediately profitable, the town passed into other hands and, being regularly laid out and surveyed, received its present title of *Empire City*. The town, thus started, acquired a mushroom growth.

During the year 1861 there was very little development of mining interests, but a general scramble for claims. The Recorder's Office was the central and only point of attraction in town. Mining claims took the place of money as an article of trade and barter. In the fall of the year, two observing and experienced miners, after careful prospecting, commended sluice mining in what are known as the Patch Diggings, occupying the slope of the hill on which the lodes were previously discovered. Meeting with unexpected success, they secured such claims as suited their purpose, and then, at the close of the season made known the secret of their success. With this followed an eager location of all adjoining claims and the subsequent development of a successful and enlarged mining interest that has made steady progress, resulting in the production of a superior quality of gold and realizing, in many instances, by the sale of claims, sudden fortunes to their lucky possessors.

Still, it is the extraneous rather than the intrinsic merits that render this place attractive to the casual traveler. A location so elevated that only the hardiest kinds of vegetation can be successfully cultivated, will never prove permanently attractive to a people accustomed to look to the earth that surrounds their homes for a supply of nutritious food. The occasional severity of the winter seasons will prove repulsive to many who enjoy its delightful summer climate. Hence there will be more or less of a fluctuating population composed largely of those who care little for home comforts or love the excitements of change of scene. To the lover of nature, however, in all its varied aspects, this point will always prove especially attractive, comprising as it does all the varieties of alpine scenery and productions, both animal and vegetable, including some of the most magnificent views to be found on the continent, taking in snowy peaks easily accessible and rivaling the highest in this part of the continental range. With the extension of roads farther west, including a country now known to be valuable in its grazing and agricultural capacities, we may expect in due course of time that Empire City will not be altogether a misnomer or that so grand a title will, by contrast with its outside appearance, suggest only a derisive smile.

Parry's Colorado Expedition of 1861 2

On October 16, 1860, Torrey wrote Gray (Rodgers 1942, pp. 278–279): "Parry writes to me that he is thinking seriously of visiting the Rocky M[oun]t[ain]s, or Pike's Peak next season—& he asks whether there is any probability of his procuring about 10 subscribers for the plants he may collect. He can—& if instructed to do so—will make excellent specimens."

During the summer of 1861, C. C. Parry collected plant specimens in the Clear Creek Valley of Colorado up to Berthoud Pass. These were reported by Asa Gray (1862–1863). Gray's enumeration was prefaced by the following pages (C. C. Parry 1862) describing the physiography and vegetation of the area.

> With the exception of a few isolated peaks and elevated ridges in connection with the Appalachian mountain range, in no instance reaching an elevation of 7,000 feet above the sea level, the truly alpine vegetation of the North American continent is confined to the remote region of the Rocky Mountains. Here alone, within temperate latitudes, do we meet with mountain ranges where the summer sun is reflected from snowy wastes, and in which occur peaks attaining an elevation of over 12,000 feet.
>
> Our previous knowledge of the general external features and peculiar vegetation of this alpine district has been derived from the researches of various explorers who have traveled hastily over this heretofore inhospitable region, noting the most prominent features of scenery along the ordinary routes of travel, determining the latitude and longitude of various fixed points, mapping out the direction of watercourses, sketching in the more prominent mountain ranges, and rarely (as in the case of James, Douglas, Drummond, Nuttall, and Frémont), making collections of its plants. From all these different sources of information, extending through the present century, we have derived a

considerable though still imperfect knowledge of the peculiar natural features of our "American Switzerland."

Within the past few years, however, the discovery of gold deposits in this portion of the mountain range has attracted thither an adventurous and enterprising population, settling with wonderful celerity its picturesque valleys and introducing into its wild recesses many of the arts and comforts of civilized life. These various social movements have afforded facilities for the prosecution of researches in natural history which were not enjoyed by the earlier pioneer explorers of this region.

In order to improve this opportunity, the writer was induced to make a journey to this region during the past season (1861), with the special object of studying its alpine vegetation and making collections of its native plants. With this view a station was selected near the foot of the dividing ridge, at the head waters of South Clear Creek. From this point an intensive scope of alpine exposure was brought within the range of an ordinary day's journey. Here, among the pine-wooded area on both sides of the Snowy Range, coursing along its alpine brooks, clambering over its precipitous rocks, floundering through snow-drifts, and mounting to its irregular crests, was spent most of the summer months of 1861. The scientific results of the observations here made, are presented in the following brief sketch and the accompanying list of plants.

The first impression made upon the traveller in approaching the mountain barrier from the broad undulating slope of the Great Plains is the irregularity of outline and apparent want of system in the grouping and arrangement of the different ridges that compose the general mass of the mountain range. Some of the higher peaks rear their snowy summits at considerable distances from the dividing crest and are met with at irregular points along the eastern slope. Numerous cross ridges interrupt the general parallelism of the principal ranges, and the actual "divide" is mostly obscured from view by elevated projecting spurs. The streams with their impetuous currents foaming along their rocky channels descend in a zigzag course, making the passage through intervening ridges by deep precipitous chasms. On reaching the more elevated mountain district the valleys become more open, and frequently spread out into oval-shaped basins, to which the name of "bars" has been applied by the miners. Towards the headwaters of the various streams these basin-shaped portions of the principal valleys, beset with scattered groves of pine, encircled by steep ridges, generally clothed with heavy growths of spruce or exhibiting occasionally smooth grassy slopes, are known as "parks." These are the miniature representatives of those larger open stretches of country which occur at the headwaters of the Platte and Grand [Colorado] rivers, forming North, South, and Middle Parks.

In approaching the dividing ridge, by following up any of the principal streams by which the mountain range is penetrated, the open parks give place to narrow valleys, generally heavily timbered with pine and spruce. The watercourses force their way through narrow rocky cañons, or, obstructed by beaver dams, spread out into marshes occupied by a tangled growth of willow and alder bushes.

The smaller tributaries which collect the waters that trickle from alpine snows ebb and flow with the diurnal changes of temperature, increasing in volume as the sun ascends to relax the icy bonds of a protracted winter, and again contracting as the clear night once more asserts the reign of perpetual frost. These alpine brooks constitute one of the most attractive features of Rocky Mountain scenery, and along their borders grow some of the finest plants of the region. Their course is that of a continuous torrent, presenting in their rapid descent a perpetual sheet of foam, rivalling in whiteness the snows in which they have their sources. Their waters of crystal purity and delicious coolness glisten in the deep shade of overhanging pines, and moisten with their spray such choice plants as *Mertensia sibirica* [*ciliata*], *Cardamine cordifolia, Saxifraga aestivalis* [*Micranthes odontoloma*], and a most elegant and conspicuous *Primula* (311) near *P. nivalis* [*parryi*].

In mounting up the steep ridges which border their course, to reach their alpine sources, the view of the surrounding country is entirely shut in by the heavy growth of pines, including on the higher ridges and abrupt slopes *Pinus contorta,* with its slender tapering trunk and stiff scanty foliage; while on more level spots, or occupying depressed basins forming subalpine marshes, *Abies alba* [*Picea engelmannii*] and *Abies balsamea* [*bifolia*] shoot up their tapering spires. The usual undergrowth in these pine woods is composed of *Vaccinium myrtillus, Shepherdia argentea* [*canadensis*], *Berberis aquifolium* [*Mahonia repens*], *Pachystima myrsinites,* etc.

In moist springy places and along the borders of marshes we find *Gaultheria myrsinites* [*humifusa*], *Pedicularis surrecta* [*groenlandica*], *Senecio triangularis, Mitella pentandra, Habenaria* [*Limnorchis*] *dilatata, Pyrola rotundifolia* var. *uliginosa,* etc. As a rarity, in scattered localities, we here meet with the charming *Calypso borealis* [*bulbosa*].

On approaching the limits of arborescent growth, indicated at first by a stunted appearance of the common varieties of pine, as well as the more frequent occurrence of the alpine species, *Pinus flexilis,* we at length come somewhat abruptly upon open stretches characterized by their peculiar vegetation and general aspect as truly alpine. Some few trees straggle for a variable distance up the abrupt rocky slopes, but in these situations they plainly exhibit the severity of the exposure by deformed and blasted trunks, often nearly prostrate, and showing by a uniform bending of their upper branches the direction of prevailing fierce winds and the weight of wintry snows. These arctic forms are confined almost exclusively to a single species of pine, heretofore undescribed (*Pinus aristata* Engelm.) belonging to the same group as *Pinus flexilis* James.

Beyond this there is a succession of alpine exposures characterized by extensive patches of snow scattered irregularly over the mountain slopes, generally indicating the accumulation of drifts, being most abundant and persistent in recesses near the higher elevations. At other points, a rough talus of rocks is spread over the surface, the separate blocks being of every conceivable shape and loosely aggregated, forming numerous fissures. In these burrowing recesses the Siberian squirrel [pika] finds a congenial abode, and salutes the traveller with his reiterated bark, often the only animate sound to break the solitude of these alpine deserts. Through these loose masses quarried out by nature's hand we often hear beneath our feet the gurgling of invisible streams, connecting by these subterranean channels elevated snowbanks with lower alpine brooks. Among these rock crevices we meet with many of the rare and attractive plants of this district, including *Aquilegia brevifolia* [*saximontana*], *Viola biflora,* a variety of *Ribes lacustre* [*R. montigenum*], *Senecio fremontii, Oxyria reniformis* [*digyna*], *Polygonum bistorta* [*Bistorta bistortoides*], etc.

Other portions of these mountain slopes are covered with a sward of alpine grasses, mingled with *Carices* and mountain clovers, all characterized by their tough, matted, and penetrating roots. In connection with these, almost every square yard presents a botanical feast of the most attractive and varied features. Neat little tufted plants of the most cerulean blue, including *Polemonium pulcherrimum* [*viscosum*], *Mertensia alpina* [*lanceolata*], *Myosotis nana* (*Eritrichum aretioides*) spot the surface. In scattered localities the bright yellow disk of *Actinella* [*Rydbergia*] *grandiflora* is conspicuous, while the varieties of alpine *Phlox, Primula angustifolia, Trifolium parryi,* etc., supply almost every tint to complete a floral rainbow. Here also by a close inspection we discover such tiny plants as *Thalictrum alpinum, Gentiana* [*Chondrophylla*] *prostrata,* and others almost hidden in the confused mass of matted foliage. In moist depressed places and along the spongy margins of alpine lakes we meet constantly with an alpine *Salix, Caltha* [*Psychrophila*] *leptosepala,* and a white *Trollius* [*albiflorus*] near *americanus.*

Toward the summit of the dividing ridge we find plants whose names plainly indicate the frigid climate to which they belong. Here grows the elegant *Claytonia* which I have called *megarhiza,* sending its deep tap-roots into the crevices of rocks whose projecting angles shelter its succulent foliage from the rude blasts that sweep over these bald exposures. Affecting similar situations we meet with an alpine *Synthyris* [*Besseya alpina*] (255) with its glossy foliage and neat spike of pale blue flowers.

On the summit of the crest, which here presents a flattened irregular surface composed of weather-worn rocks imbedded in the coarse debris of its disintegrating granitic masses, we find *Trifolium nanum, Stenotus pygmaeus, Papaver nudicaule* [*kluanense*], *Saxifraga* [*Hirculus*] *serpyllifolia, Gentiana* [*Gentianodes*] *frigida,* and others, all indicative of a rigorous climate, whose brief summer is thus elegantly adorned by these arctic forms of vegetation. Among the rarities of this district we may notice the newly discovered (or rediscovered) *Chionophila* [*jamesii*] (256), *Pedicularis sudetica* [*scopulorum*], and several others well known in the Old World, but now for the first time added to the North American flora.

Such is a general and very imperfect sketch of the prominent features of the vegetation belonging to this elevated district of *Clear Creek,* to which from my frequent visits I involuntarily applied the name of *Mount Flora.*

In my solitary wanderings over these rugged rocks and through these alpine meadows, resting at noon-day in some sunny nooks overlooking wastes of snow and crystal lakes girdled with midsummer ice, I naturally associated some of the more prominent mountain peaks with distant and valued friends. To two twin peaks always conspicuous whenever a sufficient elevation was attained, I applied the names of Torrey and Gray; to an associated peak, a little less elevated but in other respects quite as remarkable in its peculiar situation and alpine features, I applied the name of Mount Engelmann. Thus, following the example of the early and intrepid botanical explorer, Douglas, I have endeavored to commemorate the joint scientific services of our triad of North American botanists by giving their honored names to three snow-capped peaks in the Rocky Mountains. With such innocent scientific pleasantry I felt at liberty to amuse the solitary hours of my mountain excursions, often wearied, but always enjoying with the keenest zest the magnificent scenery and rich botanical treasures that lay scattered along my varied path.

Mount Engelmann from Stanley Mountain, 1982. *Photo by David Hill.*

No description indeed can do justice to the grand features of scenery brought to view from the elevated points and commanding crests of this broad mountain range. While to the east the comparatively level plain stretches out like a boundless sea; in every other direction rise elevated peaks and snow-girt ridges, stemming in deeply sheltered valleys. An obscure parallelism of the principal ridges is here for the first time noticeable, more evidently marked, however, by the occurrence of culminating points forming broken lines extending northwest and southeast than by any continuity of the principal ridges. The watershed itself is a very irregular line, difficult to trace with the eye even from the most elevated points. This is owing to a very marked peculiarity of the range which exhibits the higher culminating points disposed quite constantly on the eastern slope of the divide, with which they are generally connected by depressed spurs. It is from these offsetting peaks that the most comprehensive views are obtained, and the general topography of the range can be best studied.

It may be noticed also that the most feasible passes over the Snowy Range are met with where the dividing ridge is inclined to an east and west course. In such situations the streams flowing thence north and south, respectively, have their sources in the most depressed portions of the range usually only a short distance apart. In such a position, near the headwaters of South Clear Creek is found the depression known as

Berthoud's Pass, discovered by an engineer of that name while engaged in making a reconnaissance for the location of a direct road from Denver to Salt Lake. On this pass the elevation at the highest point does not reach above the limits of arborescent growth, the dividing waters on either side heading but a few feet apart, in a pine [spruce] grove.

Further observation will be required to show how far the accumulated snows of winter may offer obstructions to a through route, accessible at all seasons. The practical difficulties interposed by the steep ascent of the main abrupt slope can no doubt be readily overcome by the construction of embankments and zigzags. When the principal height is once gained, further progress is easy in either direction by the usual appliances of road construction, for which the proper materials of stone and lumber are abundant and of excellent quality.

The westward view takes in that irregular scope of country including Middle Park, with its broad open spaces encircled by broken ranges of mountains. These mountains send down into the plain below numerous spurs, heavily timbered with a magnificent growth of spruce (*Abies alba* [*Picea engelmannii*]). Between these ridges, deeply sheltered valleys collect the tributary streams, forming the headwaters of Grand [Colorado] River. The projecting mountain peaks on this side do not attain the height of those met with on the eastern slope, but the general surface is more elevated; the lowest depressions, occurring in the basin of Middle Park, being considerably higher than corresponding points on the Great Plains to the eastward. Hence the streams are less rapid, and the vegetation indicates a colder and more humid climate.

Here during the rainy season, in the months of July and August, the different surface exposures give rise to variable atmospheric currents which, meeting at various points, occasion a rapid development of clouds and aqueous precipitation such as characterizes the sudden showers in this peculiar district. Here in fact may be studied to the best advantage (though not always under agreeable circumstances) the formation of clouds in their endless variety of shape, density, and progressive development. These at times may be seen gradually accumulating about the summits of snow covered peaks, thence spreading over the horizon and extending to the zenith, causing a regular steady rain; while at other times a sudden gust calls attention to a rapidly forming angry cloud, which sweeps over the surface in a well defined path, scattering rain, hail, or snow in its wake.

The regular afternoon showers which occur on the eastern slope are readily explained by referring them to the junction of heated air, charged with moisture, ascending from the Great Plains, with the descending currents of cold air from the Snowy Range, by which the moisture of the former is precipitated. As soon as the equilibrium is established, the rain passes off and a sky more or less clear succeeds, followed almost invariably by clear nights and bright mornings. This series of phenomena, often continuing with remarkable regularity from one day to another, continues through the months of July and August, constituting a rainy season.

The principal object of my journey being the collection of plants, I may here very properly conclude this sketch of the general features of scenery and climate. The accompanying list of plants prepared from my collections, and notes by Prof. Gray and Dr. Engelmann, will serve to give a more precise view of the botany of this region, particularly of the alpine district, to which my attention was specially directed. Travelling over a path so ably investigated by early explorers, I have still been rewarded for my labors by the discovery of several interesting novelties, as well as by adding quite a number of alpine plants, well known in the Old World, to our North American Flora.

Should circumstances prove favorable, it is the intention of the writer to continue these observations during the coming season, over a wider section of country lying to the west and south of the investigations of the past season.

❧

The following is a list of the plants, as given in Gray (1862–1863), minus the often copious notes that are primarily of interest to readers concerned with the evolution of taxonomic concepts.

List Of Plants

Parry's collection lacks some numbers cited by Gray. Possibly these were unicates and sent to Gray, perhaps some are listed among the unnumbered collections, or perhaps Parry's personal collections were used to augment the specimens distributed a year later by Hall & Harbour. Some numbers were not seen by Gray although they are found in Parry's collection.

Unfortunately, Parry's labels are usually without any specific locality data, and his numbers are sometimes switched or unreliable in the duplicate sets. It also appears that when Parry ran out of 1861 printed labels he used labels of 1862, 1864, or 1872. When the collection number traces back to Gray's list it is clear that the year of collection was 1861. Occasionally the number would be lacking from a specimen; this could be recovered by consulting Gray's list.

There is sometimes a problem of deciding whether a particular Parry specimen of a taxon described by Gray from Parry's 1861 plants actually is the type collection. I have followed the principle that if the specimen is the only one of the species in Parry's collection, there is a very high likelihood that it is the isotype, even if there is no number on the label or if Parry used a blank label from another collecting year. For all of his erudition in his published work, Parry seems to have been exceptionally unreliable in his collecting data, or rather lack of it.

The enumerations of plants that are given for each year of Parry's visits to Colorado contain some standard botanical shorthand that might not be understood by the lay reader.

1. Standard international abbreviations for herbaria: ISC stands for Iowa State University; DUKE for Duke University; G for Jardin Botanique, Geneva, Switzerland; GH for Gray Herbarium, Harvard University; MO for Missouri Botanical Garden; COLO for University of Colorado.
2. The name given to a specimen at GH by Gray (1862–1863) is given in bold-face if it has not been changed, in italics if it now goes under another name in Weber and Wittmann, *Catalog of the Colorado Flora* (1992).
3. An exclamation point preceding a herbarium abbreviation means that I have examined the specimen at that herbarium.
4. The text in square brackets represents my own comments.
5. Author names are omitted because the complete authorities can readily be located in Weber and Wittmann (1992). This practice is recommended for nontechnical papers by Garnock-Jones et al. (1996).
6. A type specimen (or holotype) is that single specimen to which the original description of the species is linked. An isotype is a duplicate specimen of the type collection. A syntype is a specimen cited with the holotype.

The plant list is as follows.

1. *Erigeron grandiflorum* var. *elatius,* isotype: **E. elatior,** !DUKE, ISC
2. **Arnica cordifolia,** !DUKE
3. "Varieties of the last [referring to no. 8]: one with blue, the other with nearly white rays, far less pubescent" [probably **Erigeron simplex,** but other possibilities exist]

4. *Erigeron macranthum:* **E. speciosus** var. **macranthus,** !ISC
5. **E. compositus** var. **discoideus,** !DUKE
6. **E. compositus,** !DUKE
7. *E. acre:* **Trimorpha elongata,** !DUKE [as no. 3]
8. *E. uniflorum:* **E. melanocephalus,** !DUKE, ISC; **E. grandiflorus,** !ISC
9. **E. bellidiastrum,** !ISC, "Bijou Creek, sandy bottom [east of Denver]"
10. *Arnica angustifolia* var. *discoidea! latifolia:* **A. parryi.** Syntype of *A. angustifolia* var. *eradiata,* !ISC
11. **Erigeron compositus**
12. *Boltonia latisquama,* isotype: **Boltonia asteroides** subsp. **latisquama**, !ISC [Despite the printed label heading "Rocky Mountain Flora, from the headwaters of Clear Creek [etc.]," this was not collected in Colorado but, as Gray stated in the protologue, "near the mouth of the Kansas River, Sept." This locality is at Kansas City, Kansas, where the river enters the Missouri.]
13. *Aster glaucus:* **Eucephalus glaucus,** !DUKE, ISC
14. *Machaeranthera canescens:* **M. pattersonii,** !ISC
15. **Solidago missouriensis,** !DUKE, ISC
16. *S. humilis:* **S. simplex,** !DUKE, ISC
17. *S. missouriensis:* **S. simplex,** !DUKE
18. *S. humilis* var. *alpina,* "only an inch or two high." **S. simplex** var. **nana**
19. *Senecio aureus* var. *balsamitae,* with leaves more pinnatifid: **Packera paupercula** [This occurs in South Park.]
20. *S. canus:* **Packera cana,** !DUKE, ISC
21. *S. lugens:* **S. wootonii,** !ISC
22. *S. canus,* "the same species [referring to no. 20] with more numerous and smaller heads" [possibly *Packera werneriifolia*]
23. *S. exaltatus* var. *minor:* **S. atratus,** !DUKE, ISC
24. **S. integerrimus**
25. **S. triangularis,** !DUKE, ISC
26. **S. eremophilus** [subsp. **kingii**], !DUKE, ISC
27. **S. fremontii** [subsp. **blitoides**], !ISC
28. **S. fremontii,**!DUKE [The sheet, !ISC, contains two specimens, one being **Ligularia holmii,** the other **L. taraxacoides,** isotype of *Senecio amplectens* var. *taraxacoides. Ligularia, sensu* Weber, is probably a distinct undescribed genus.]
29. *Palafoxia hookeriana:* **P. sphacelata** [collected on the plains]
30. *Aplopappus spinulosus:* **Machaeranthera pinnatifida** [collected on the plains]

31. *Coreopsis involucrata:* **Bidens polylepis,** !ISC [Probably collected in Kansas or Nebraska. The label Parry used is misleading. There is no other evidence of the species occurring in Colorado. McGregor et al. (1977) do not report it west of eastern Nebraska and Kansas.]
32. *Arnica angustifolia:* **A. chamissonis** subsp. **foliosa,** !ISC
33. *Townsendia sericea:* **T. hookeri**, !DUKE, ISC [as no. 35]. Also reported as *Erigeron compositus*: **E. pinnatisectus,** !ISC
34. *Cirsium edule,* "a common subalpine species, flowers yellowish": **Cirsium parryi**
35. *Townsendia sericea:* **T. hookeri**
36. *Euphrosyne xanthifolia:* **Cyclachaena xanthifolia,** !ISC, "Platte Valley"
37. *Antennaria dioica* var. *rosea:* **A. rosea**
38. *A. carpathica:* **A. pulcherrima** subsp. **anaphaloides,** !ISC
39. *A. dioica:* **A. rosea,** !ISC [Label lacks a number.]
40. **Iva axillaris,** !DUKE, ISC, "8 Mile Creek, [Fremont Co.], May 30"
41. *Artemisia borealis:* **Oligosporus groenlandicus,** !DUKE
42. *A. richardsoniana*: **A. arctica** subsp. **saxicola,** !DUKE
43. **A. frigida,** !DUKE [as no. 45]
44. *A. filifolia:* **Oligosporus filifolius**
45. *A. canadensis:* **Oligosporus pacificus,** !DUKE [as no. 43]
46. *Actinella acaulis:* **Tetraneuris brevifolia,** !DUKE, ISC
47. *Aplopappus (Stenotus) pygmaeus:* **Tonestus pygmaeus,** !DUKE, ISC
48. *Grindelia squarrosa:* **G. subalpina,** !DUKE, ISC
49. *Linosyris viscidiflora:* **Chrysothamnus viscidiflorus,** !DUKE, ISC
50. **Helianthus pumilus,** !ISC
51. *Aplopappus parryi*, isotype: **Oreochrysum parryi**
52. *Senecio cernuus*, isotype: **Ligularia pudica,** !DUKE, ISC
53. **Arnica mollis**
54. *A. angustifolia:* **A. rydbergii,** !DUKE, ISC
55. *Chaenactis achilleaefolia:* **C. alpina,** !DUKE, ISC
56. *Senecio amplectens*, isotype: **Ligularia amplectens,** !ISC
57. *Helianthus orgyalis:* **H. salicifolius** [This was not collected in Colorado, more likely in eastern Kansas, "mouth of Kansas River," along with *Boltonia asteroides*. This specimen is probably the basis of Rydberg's (1932) mention of "E. Colorado."]
58. *Villanova chrysanthemoides:* **Bahia dissecta,** !DUKE, ISC
59. *Chrysopsis villosa:* **Heterotheca villosa,** !DUKE
60. *Aplopappus (Stenotus) pygmaeus:* **Tonestus pygmaeus**
61. *Actinella grandiflora:* **Rydbergia grandiflora,** !DUKE, ISC
62. **Gaillardia aristata,** !DUKE [as no. 65], ISC

63. *Senecio aureus* var. *alpinus,* isotype: **Packera werneriifolia,** !ISC
64. *Macrorhynchus troximoides:* **Agoseris aurantiaca**
65. *Troximon glaucum:* **Agoseris glauca,** !ISC
66. *Macrorhynchus troximoides:* **Agoseris aurantiaca,** !ISC
67. *Troximon parviflorum:* **Agoseris glauca,** !ISC
68. **Lygodesmia juncea,** !DUKE, ISC
69. *Crepis runcinata:* **Psilochenia occidentalis,** !ISC
70. No entry. *Crepis occidentalis:* **Psilochenia occidentalis,** !ISC
71. *Hieracium fendleri:* **Chlorocrepis fendleri,** !DUKE, ISC
72. *H. triste:* **Chlorocrepis tristis** subsp. **gracilis,** !DUKE, ISC
73. *Mulgedium pulchellum:* **Lactuca tatarica** subsp. **pulchella,** !ISC
74. *Atragene alpina* var. *ochotensis:* **A. columbiana,** !ISC [mounted with *Hall & Harbour 1*]
75. **Thalictrum alpinum** L., !DUKE, ISC
76. **T. sparsiflorum,** !DUKE [The specimen at ISC is *T. fendleri.*]
77. *Ranunculus affinis:* **R. cardiophyllus,** !ISC [mounted with *Hall & Harbour 16*]
78. *R. cymbalaria:* **Halerpestes cymbalaria** subsp. **saximontana**
79. *R. glaberrimus:* **R. alismifolius** var. **montanus,** !DUKE, ISC [Gray corrected his identification on p. 404.]
80. **R. eschscholtzii,** !ISC [A second sheet is **R. inamoenus.**]
81. *R. amoenus?:* **R. adoneus,** isotype, !DUKE, ISC
82. *Clematis douglasii:* **Coriflora hirsutissima**
83. *Trollius laxus* var. *albiflorus,* isotype: **T. albiflorus,** !ISC
84. *Delphinium elatum:* **D. barbeyi,** !DUKE, ISC
85. *D. scopulorum:* **D. x occidentale,** !DUKE, ISC
86. *Aconitum nasutum:* **A. columbianum**
87. *Anemone multifida:* **A. multifida** var. **globosa,** !DUKE, ISC
88. *Pulsatilla nuttalliana:* **P. patens** subsp. **multifida**
89. **Aquilegia coerulea**
90. *A. vulgaris* var. *brevistyla:* **A. saximontana**
91. *Caltha leptosepala:* **Psychrophila leptosepala,** !ISC
92. *Thlaspi cochleariforme:* **Noccaea montana** (as *T. fendleri, T. alpestre*), !ISC
93. *Draba johannis:* **D. crassifolia,** !ISC [mounted with *Hall & Harbour 41*]
94. *Turritis patula:* **Boechera sp.**
95. *Erysimum pumilum:* **E. capitatum** [the alpine ecotype]
96. **Draba streptocarpa**, isotype, !DUKE, ISC [From upper Clear Creek. Unnumbered; the printed label "1862" is probably incorrect, for this is the only specimen of the species in Parry's collection.]

97. **D. nemorosa**
98. **Arabis hirsuta** [var. **pycnocarpa**], !DUKE, ISC
99. **Cardamine cordifolia,** !DUKE, ISC
100. *Sisymbrium canescens:* **Descurainia sp.**
101. *Physaria didymocarpa:* **P. vitulifera,** !ISC [mounted with *Hall & Harbour 47*]
102. **Erysimum asperum** [collected on the plains]
103. **Draba aurea,** !ISC [mounted with *Hall & Harbour 44*]
104. *Cleomella tenuifolia:* **C. angustifolia**
105. *Cleome integrifolia:* **C. serrulata**
106. **Viola biflora**
107. *V. muhlenbergii:* **V. adunca,** !ISC
108. *V. muhlenbergii:* **V. adunca**
109. **V. nuttallii,** !ISC [from the plains]
110. *V. palustris:* **V. macloskeyi** subsp. **pallens**
111. *Geranium carolinianum:* **G. bicknellii** var. **longipes,** !ISC [This taxon has been considered an alien in Colorado, but its occurrence here as early as 1861 suggests that it is native.]
112. **G. richardsonii,** !ISC
113. *G. fremontii* var. *parryi,* isotype: **G. caespitosum**
114. **Gaura coccinea,** !DUKE, ISC
115. *Oenothera lavandulaefolia:* **Calylophus lavandulifolius,** !ISC
116. **O. albicaulis,** !DUKE
117. *O. albicaulis,* "the same, with undivided leaves": **O. latifolia,** !DUKE, ISC
118. *Stenosiphon virgatus:* **S. linifolium** [Collected on the plains, probably not in Colorado, although this is now known to occur in southeastern Baca County.]
119. *Epilobium tetragonum:* **E. cf. saximontanum,** !ISC
120. *E. alpinum:* **E. anagallidifolium,** !DUKE
121. *E. alsinifolium* [Not an American species. Perhaps **E. hornemannii.**]
122. *E. tetragonum*: **E. cf. halleanum,** !ISC
123. *E. latifolium:* **Chamerion subdentatum,** !DUKE
124. **Gayophytum ramosissimum: G. ramosissimum + G. diffusum,** !DUKE, ISC [*E. diffusum* only]
125. *Epilobium paniculatum:* **Epilobium brachycarpum,** !DUKE, ISC
126. *Mentzelia albicaulis:* **Acrolasia albicaulis,** !ISC
127. *M. nuda:* **Nuttallia nuda,** !ISC
128. *Sedum rhodiola:* **Rhodiola integrifolia,** !ISC
129. *Sedum rhodanthum,* isotype: **Clementsia rhodantha,** !ISC

130. *S. stenopetalum:* **Amerosedum lanceolatum,** !ISC
131. *Silene drummondii:* **Gastrolychnis drummondii**
132. *Lychnis apetala:* **Gastrolychnis drummondii,** !ISC
133. *L. apetala:* **Gastrolychnis drummondii,** !ISC
134. **Silene scouleri** [subsp. **hallii**], !ISC
135. *Gentiana humilis:* **Chondrophylla aquatica,** !ISC
136. **Stellaria longifolia,** !DUKE, ISC
137. *Silene menziesii:* **Anotites menziesii,** !DUKE, ISC
138. *Cerastium vulgatum* var. *beeringianum,* and *C. arvense:* **C. beeringianum** subsp. **earlei,** !ISC, and **C. strictum**, !ISC
139. *Sagina linnaei:* **S. saginoides,** !DUKE, ISC
140. *Arenaria fendleri:* **Eremogone fendleri,** !DUKE, ISC
141. *A. arctica* var.: **Lidia obtusiloba,** !DUKE
142. *Claytonia arctica* var. *megarhiza,* isotype: **Claytonia megarhiza,** !ISC [also unnumbered, with label for 1862]
143. *Talinum pygmaeum,* isotype: **Oreobroma pygmaea,** !ISC
144. **Ceanothus fendleri,** !DUKE
145. **C. velutinus,** !ISC
146. *Berberis aquifolium* var. *repens:* **Mahonia repens**
147. *Papaver alpinum:* **Papaver kluanense,** !ISC
148. **Callirhoë involucrata,** !ISC [collected on the plains]
149. *Ribes lacustre,* "an alpine form": **R. montigenum,** !ISC
150. **R. cereum,** !ISC
151. *R. hirtellum:* **R. inerme,** !ISC
152. *R. prostratum:* **R. coloradense,** !ISC
153. *Rhus trilobata,* a variety of *R. aromatica:* **Rhus aromatica** subsp. **trilobata,** !ISC
154. *Archangelica gmelinii:* **Angelica grayi,** !ISC
155. *Berula angustifolia:* **Oxypolis fendleri,** !ISC
156. *Conioselinum fischeri:* **C. scopulorum,** !ISC [In the published paper, an unnumbered reference is made to a collection of *Leptotaenia dissecta* [**Lomatium dissectum**] from the foot of the mountains. This is an unlikely find in the area covered and may refer to *Ligusticum porteri.*]
157. *Cymopterus terebinthinus* var. *foeniculaceus:* **Aletes anisatus,** !ISC [labeled *Cymopterus? anisatus*]
158. *C. alpinus,* isotype: **Oreoxis alpina,** !ISC
159. *Thaspium montanum* var. *tenuifolium:* **Pseudocymopterus montanus**
160. *Cymopterus montanus:* **C. acaulis,** !ISC, "Upper Platte, May"
161. *cf. Thaspium montanum,* "in flower only": **Pseudocymopterus montanus**

162. **Paxistima myrsinites,** !ISC
163. *Saxifraga punctata:* **Micranthes odontoloma,** !DUKE, ISC
164. *S. hirculus:* **Hirculus prorepens,** !DUKE, ISC [The specimen at ISC is missing, noted as "probably having been sent to Gray."]
165. *S. flagellaris:* **Hirculus platysepalus** subsp. **crandallii,** !ISC
166. *S. hirculus,* "a small form, only 2 or 3 inches high, but quite like the common Arctic American specimens": **Hirculus serpyllifolius** subsp. **chrysanthus,** !DUKE, ISC [The specimen missing, noted as "probably sent to Gray"; the type of *S. chrysantha,* however, was *Hall & Harbour 199.*]
167. **S. cernua,** !DUKE, ISC
168. *S. bronchialis:* **Ciliaria austromontana,** !DUKE, ISC
169. *S. nivalis:* **Micranthes rhomboidea,** !ISC
170. *S. caespitosa* var.: **Muscaria delicatula,** !DUKE, ISC
171. **Mitella pentandra,** !DUKE, ISC
172. **Heuchera bracteata,** !DUKE, ISC
173. **H. parvifolia,** "strictly alpine," !DUKE, ISC [Labels switched between this and the next.]
174. *H. parvifolia:* **H. parvifolia** var. **nivalis,** !DUKE, ISC
175. **Jamesia americana,** !ISC
176. **Trifolium dasyphyllum,** !DUKE, ISC
177. **T. nanum,** !DUKE, ISC
178. **T. parryi,** isotype, !DUKE, ISC
179. **Oxytropis splendens,** !ISC
180. *Phaca oroboides:* **Astragalus eucosmus,** !ISC
181. *Astragalus nigrescens:* **A. tenellus,** !DUKE, ISC
182. **A. alpinus,** !DUKE, ISC
183. **Oxytropis lambertii** and **O. sericea,** !ISC [**O. sericea** only]
184. *Astragalus cf. glareosus:* **A. mollissimus,** !ISC
185. **A. pectinatus,** !ISC [on the plains]
186. **Oxytropis lambertii,** !ISC
187. *Lathyrus ornatus,* isotype of var. *glabratus:* **L. polymorphus,** !ISC, "on the lower Platte"
188. *L. linearis:* **Vicia americana** var. **minor,** !DUKE, ISC
189. **Astragalus gracilis,** syntype of *A. microlobus,* !ISC [as no. 189]
190. *A. sericoleucus:* **Orophaca sericea,** !ISC, "Gravelly hills, Upper Platte"
191. *Oxytropis nana:* **O. multiceps,** !DUKE, ISC
192. *Dalea alopecuroides:* **D. leporina,** !ISC
193. **Astragalus parryi,** isotype, !ISC [Parry used an 1862 label for this, but it is obvious from the publication date and Gray's inclusion in this list that the plant was collected in 1861.]

194. *Hosackia purshiana:* **Lotus wrightii,** !DUKE, ISC, "Valley of the Platte"
195. *Dalea laxiflora:* **D. enneandra,** !DUKE, ISC
196. *Sophora sericea:* **Vexibia nuttalliana,** !ISC
197. *Thermopsis rhombifolia:* probably **T. divaricarpa,** !ISC [It is unlikely that Parry arrived in time to find the very early-flowering *T. rhombifolia.*]
198. *Psoralea lanceolata:* **Psoralidium lanceolatum,** !ISC
199. **Oxytropis lambertii,** !DUKE, ISC [incorrectly given by Gray as 189]
200. *Lupinus sp.:* **L. argenteus,** !ISC
201. *Prunus virginiana:* **Padus virginiana** subsp. **melanocarpa,** !ISC
202. **Sibbaldia procumbens,** !ISC
203. **Dryas octopetala** [subsp. **hookeriana**], !DUKE, ISC
204. **Geum rivale,** !DUKE, ISC
205. *G. rossii:* **Acomastylis rossii** subsp. **turbinata,** !DUKE, ISC
206. *Spiraea discolor:* **Holodiscus discolor,** !DUKE, ISC
207. *S. opulifolia:* **Physocarpus monogynus,** !ISC
208. *Rosa blanda:* mixture, **R. woodsii** and **R. arkansana,** !ISC
209. *Cercocarpus parvifolius:* **C. montanus,** !ISC
210. *Rubus deliciosus:* **Oreobatus deliciosus,** !DUKE, ISC
211. *R. nutkanus:* **Rubacer parviflorum,** !ISC
212. **Rubus idaeus** [subsp. **melanolasius**], !ISC
213. *Potentilla fissa:* **Drymocallis fissa**
214. **P. nivea**
215. **P. nivea,** !DUKE

216–217. **P. pensylvanica** var. **strigosa**: 217 is **P. pulcherrima,** !DUKE

218–220. **P. diversifolia:** 219 is **P. subjuga,** !DUKE

221. **Adoxa moschatellina,** !ISC
222. *Sambucus racemosa:* **S. microbotrys,** !ISC
223. *Symphoricarpos montanus; S. oreophilus:* **S. rotundifolius,** !ISC
224. *Lonicera involucrata:* **Distegia involucrata,** !ISC
225. *Viburnum pauciflorum:* **V. edule,** !ISC
226. **Vaccinium cespitosum,** !DUKE, ISC
227. *V. myrtillus* var. *microphyllum:* **V. scoparium,** !DUKE, ISC
228. *V. myrtillus* var.?: **V. scoparium,** !ISC
229. **Pyrola minor,** !DUKE, ISC
230. **P. chlorantha,** !DUKE, ISC
231. *P. uniflora:* **Moneses uniflora,** !DUKE, ISC [unnumbered]
232. **P. rotundifolia** var. *uliginosa:* subsp. **asarifolia,** !DUKE, ISC
233. *P. secunda:* **Orthilia secunda,** !DUKE, ISC

234. *Gaultheria myrsinites:* **G. humifusa,** !DUKE, ISC
235. *Mimulus luteus:* **M. guttatus,** !ISC
236. **Collinsia parviflora,** !ISC
237. *Veronica alpinus:* **V. nutans**
238. *Gerardia aspera:* **Agalinis tenuifolia,** !ISC, "Valley of the Platte"
239–241. *Castilleja pallida:* 239 is **C. rhexifolia,** !DUKE, ISC, "Mill Cr., June" [There is a Mill Creek in Clear Creek Co. The specimen at GH discussed by Gray is evidently **C. sulphurea**]
242. *C. pallida:* **C. occidentalis,** !ISC
243. *C. breviflora:* **C. puberula,** !DUKE, ISC
244. **C. integra,** !ISC
245. *C. pallida:* **C. rhexifolia,** !ISC [The specimen discussed by Gray is evidently *C. sulphurea*]
246. **C. linariifolia,** !DUKE, ISC
247. **Orthocarpus luteus,** !DUKE
248. **Pedicularis racemosa** [subsp. **alba**], !DUKE, ISC
249. **P. bracteosa** [subsp. **paysoniana**], !DUKE, ISC
250. **P. groenlandica,** !DUKE, ISC
251. **P. parryi,** isotype, !DUKE [as no. 255], ISC
252. **P. procera,** isotype, !DUKE, ISC
253. *P. sudetica* var.: **P. scopulorum,** syntype, !DUKE, ISC
254. *Synthyris plantaginea:* **Besseya plantaginea,** !DUKE, ISC
255. *S. alpina,* isotype: **Besseya alpina,** !DUKE, ISC
256. **Chionophila jamesii,** !DUKE, ISC
257. *Penstemon humilis:* **P. virens,** !DUKE, ISC
258. *P. acuminatus:* **P. secundiflorus,** !DUKE, ISC
259. **P. glaber** var. *alpinus:* var. **glaber**., !DUKE, ISC
260. **P. glaber,** !ISC
261, 262. *P. glaucus* var. *stenosepalus,* isotype [as to no. 261]: **P. whippleanus,** !ISC
263. *P. procerus,* !DUKE, ISC: **P. confertus** subsp. **procerus**
264. *P. acuminatus,* "a narrow leaved variety": **P. secundiflorus,** !ISC [This dwarfed form, possibly a distinct species, is common in the dry grasslands of South Park.]
265. **P. albidus,** !ISC, "Upper Platte, May"
266. *Campanula langsdorffiana:* **C. parryi,** isotype, !ISC
267. **C. uniflora,** !ISC [A second sheet bears the number 266.]
268. **C. rotundifolia,** !ISC
269. *Valeriana dioica:* **V. capitata** subsp. **acutiloba,** !ISC
270. *Galium boreale:* **G. septentrionale,** !ISC
271. *Gilia spicata:* **Ipomopsis spicata,** !ISC [sheet with three specimens,

the central one marked "271," the others probably collected, according to the labels, in Wyoming (no. 239), "gravelly hills N of Bridger, June 12, 1873" on the Northwestern Wyoming Expedition, Capt. W. A. Jones, commanding]

272. **Phacelia sericea,** !DUKE, ISC
273. *Cuscuta cuspidata:* **Grammica cuspidata,** !ISC, "Bijou Creek, Aug."
274. *Polemonium pulcherrimum:* **P. viscosum,** !DUKE, ISC
275. *P. coeruleum:* **P. foliosissimum,** isotype, !ISC
276. *P. pulchellum:* **P. pulcherrimum** subsp. **delicatum,** !DUKE, ISC
277. **Ipomoea leptophylla,** !ISC, "Sand hills of the Platte"
278. *Eritrichium villosum* var. *aretioides:* **Eritrichum aretioides,** !DUKE, ISC
279. **Primula angustifolia,** !DUKE, ISC
280. **Collomia linearis,** !ISC, "Central City, May 14"
281. *C. gracilis:* **C. linearis,** !ISC
282. **Gilia pinnatifida,** !ISC
283. *G. aggregata:* **Ipomopsis aggregata** subsp. **collina,** !DUKE
284. *Mertensia alpina:* **M. lanceolata,** *sens. lat.* (*M. viridis*), !DUKE; **M. lanceolata,** *sens. str.*, !ISC
285. *M. sibirica:* **M. ciliata,** !DUKE, ISC
286. *M. paniculata:* **M. lanceolata,** *s. lat.* (*M. viridis*), !DUKE, ISC
287. *M. alpina:* **M. lanceolata,** *s. lat.* (*M. viridis*), !ISC
288. *Eritrichium glomeratum:* **Oreocarya virgata,** !DUKE, ISC
289. *Phacelia circinata:* **P. heterophylla,** !DUKE, ISC
290. *Echinospermum floribundum:* **Hackelia floribunda**
291. *Eritrichium crassisepalum:* **Oreocarya suffruticosa,** !ISC

292–294. No entry

295. *Lithospermum pilosum, ex char.:* **L. multiflorum,** !DUKE, ISC
296. *Heliotropium convolvulaceum:* **Euploca convolvulacea,** !ISC [on the plains]
297. *Paronychia sp.:* **P. pulvinata,** !DUKE, ISC
298. *Phlox hoodii* var: *a.*, **P. hoodii,** *b.*, **P. condensata,** isotype of *P. caespitosa* var. *condensata,* !ISC
299. *Gilia pungens:* **Leptodactylon pungens,** !ISC
300. **Silene acaulis** [subsp. **subacaulescens**], !ISC
301. **Dracocephalum parviflorum,** !ISC
302. *Salvia pitcheri:* **S. azurea** var. **grandiflora,** !DUKE, ISC [on the plains]
303. *Scutellaria resinosa:* **S. brittonii,** !ISC, "upper Platte"
304. *Gentiana parryi,* isotype: **Pneumonanthe parryi,** !ISC
305. *G. frigida* var. *algida:* **Gentianodes algida,** !ISC

306. *G. prostrata* var. *americana:* **Chondrophylla prostrata**
307. *G. acuta* var. *stricta:* **Gentianella acuta,** !ISC
308. **Swertia perennis**
309. *Gentiana acuta* var. *nana:* **Comastoma tenellum**
310. **Frasera speciosa,** !ISC
311. **Primula parryi,** isotype, !DUKE
312. *Dodecatheon meadia:* **D. pulchellum,** !DUKE, ISC
313. **Androsace septentrionalis,** !ISC
314. *Phacelia popei:* **P. alba,** !DUKE, ISC
315, 316. **Eriogonum umbellatum,** !DUKE, ISC [Both numbers mounted on the same sheet. Gray's note that the GH specimen of 316, "the perianth changed to pale yellow turning brownish" suggests *E. subalpinum.*]
317. **E. flavum** [subsp. **chloranthum**], !DUKE, ISC
318. **E. umbellatum,** [The DUKE specimen, and evidently the GH specimen described by Gray as having flowers of white to cream color, is *E. subalpinum.*]
319. *E. alatum:* **Pterogonum alatum,** !DUKE, ISC
320. **E. annuum,** !ISC [collected on the plains]
321. **E. effusum,** !DUKE, ISC, "plains east of Denver"
322. *Polygonum tenue:* **P. douglasii,** !DUKE, ISC, "hillsides near Central City"
323. *Montelia tamariscina?:* **Amaranthus arenicola** [collected on the plains]
324. *Euphorbia marginata:* **Agaloma marginata,** !ISC
325. *Croton muricatum:* **C. texensis**
326. *Froelichia floridana:* **F. gracilis,** !DUKE
327. *Cycloloma platyphyllum:* **C. atriplicifolium,** !DUKE
328. *Eurotia lanata:* **Krascheninnikovia lanata,** !ISC, "plains E of Denver, Aug. 1861"
329. *Euphorbia hexagona:* **Zygophyllidium hexagonum,** !ISC [collected on the plains]
330. *E. petaloidea:* **Chamaesyce missurica,** !ISC
331. **Solanum rostratum,** !ISC, "Platte Valley" [on the plains]
332. *Polygonum viviparum:* **Bistorta vivipara,** !DUKE, ISC
333. *P. bistorta* var. *oblongifolium:* **Bistorta bistortoides,** !DUKE, ISC
334. **Oxyria digyna,** !DUKE, ISC
335. *Asclepias verticillata:* **A. pumila,** !ISC
336. *Abronia cycloptera:* **Tripterocalyx micranthus,** !DUKE
337. **A. fragrans,** !DUKE [collected on the plains]
338. **Acer glabrum,** !ISC

339. *Betula alba* var. *glutinosa:* **B. fontinalis,** !ISC
340. *Alnus viridis:* **A. incana** subsp. **tenuifolia,** !ISC
341. *Salix glauca:* **S. brachycarpa,** !DUKE, ISC
342. *S. cordata?:* **S. monticola,** !ISC
343. **S. reticulata** [subsp. **nivalis**], !ISC
344. *S. discolor:* **S. planifolia,** !ISC
345. **Populus tremuloides,** !ISC
346. **Lloydia serotina,** !DUKE, ISC
347. *Calochortus venustus:* **C. gunnisonii,** !DUKE; !ISC [unnumbered]
348. *Streptopus amplexifolius:* **S. fassettii,** !DUKE, ISC
349. **Leucocrinum montanum,** !DUKE, ISC
350. **Allium cernuum,** !DUKE, ISC
351. *Zygadenus glaucus:* **Anticlea elegans,** !DUKE, ISC
352. *Coralloriza innata:* **C. trifida,** !ISC
353. **Listera cordata** [subsp. **nephrophylla**]
354. *Calypso borealis:* **C. bulbosa,** !DUKE, ISC
355. *Platanthera obtusata:* **Lysiella obtusata,** !DUKE, ISC
356. *P. hyperborea:* **Limnorchis hyperborea,** !ISC [+ **L. stricta**]
357. *P. dilatata:* **Limnorchis dilatata** subsp. **albiflora,** !ISC
358. **Juncus castaneus,** !DUKE, ISC
359. **J. triglumis,** !DUKE
360. *J. arcticus* var. *gracilis?:* **J. parryi,** !ISC
361. *J. menziesii:* **J. longistylis,** !DUKE, ISC
362. **Luzula parviflora,** !DUKE, ISC
363. *Poa alpina?:* **Poa lettermanii,** !ISC
364. **Munroa [= Monroa] squarrosa,** !ISC [unnumbered]
365. *Calamagrostis sylvatica:* **C. purpurascens,** !ISC
366. *Muhlenbergia gracilis:* **M. montana,** !ISC [unnumbered]
367. *Aira caespitosa* var. *arctica:* **Deschampsia cespitosa,** !ISC
368. **Calamagrostis purpurascens,** !DUKE
369. **Buchloë dactyloides,** !ISC, "plains of the Platte"
370. *Bouteloua oligostachya:* **Chondrosum gracile,** !DUKE, ISC
371. *Eriocoma cuspidata:* **Achnatherum** (*Oryzopsis, Stipa*) **hymenoides,** !DUKE
372. *Aira caespitosa:* **Deschampsia cespitosa,** !ISC [unnumbered]
373. *Festuca rubra,* "too young": **F. brachyphylla** subsp. **coloradensis,** !ISC
374. *Poa laxa:* **P. glauca,** !ISC
375. *P. nemoralis:* **P. fendleriana,** !ISC
376. **P. arctica,** !DUKE, ISC
377. *Trisetum subspicatum:* **T. spicatum** subsp. **congdonii,** !DUKE, ISC

378. *Bromus kalmii:* **Bromopsis porteri,** !ISC, “S. Clear Creek”
379. *Poa andina:* **P. fendleriana,** !DUKE, ISC, “Upper Clear Creek”
380. *Festuca ovina:* **F. brachyphylla** subsp. **coloradensis,** !DUKE, ISC
381. *Triticum aegilopoides:* **Elymus scribneri,** !ISC
382. *Phleum alpinum:* **P. commutatum,** !DUKE, ISC
383. *Carex atrata* var. *nigra:* **C. sp., atrata** group
384. *C. rigida:* **C. nebrascensis**
385. *C. incurva:* **C. perglobosa** and **C. vernacula,** !ISC [printed label of 1892 used; a scrap label attached to *Hall & Harbour 142*]
386. **C. capillaris,** !DUKE, ISC
387. *C. atrata* var. *nigra:* **C. scopulorum,** !DUKE, ISC [The ISC specimen bears a printed label for 1872, but this is the only sheet in the collection.]
388. **C. aurea,** !ISC
389. *C. atrata* var. *nigra:* **C. chalciolepis, C. cf. arapahoensis, C. egglestonii** , !DUKE
390. **C. lanuginosa,** !ISC, “South Clear Creek, July”
391. *C. festiva:* **C. microptera,** !ISC, “S. Clear Creek”
392. **Luzula spicata,** !DUKE, ISC [as no. 363]
393. *Carex bromoides?* “too young”: **C. deweyana,** !ISC
394. *Woodsia obtusa:* **W. scopulina,** syntype, !ISC
395. **Cystopteris fragilis,** !ISC
396. *Allosorus acrostichoides:* **Cryptogramma acrostichoides,** !DUKE, ISC
397. *Nothochlaena dealbata:* **Argyrochosma fendleri,** !ISC, “near Idaho [Springs]”
417. *Aster sp.:* Type collection of *Aster adscendens* var. *parryi* = **A. foliaceus** var. **parryi.** Not seen.

It is entirely possible that the following unnumbered collections were from 1862 but that Parry used labels for 1861. There seems to be no reason Parry would not have sent these to Gray with numbers if they had really been collected in 1861.

Amelanchier: **A. alnifolia,** !ISC
Aphyllon fasciculatum, !ISC
Ambrosia: **A. psilostachya** var. **coronopifolia,** !ISC, “Colorado Plains”
Aplopappus radiata: **Pyrrocoma crocea,** !ISC [This is probably *Hall & Harbour 257*, the isotype of *Aplopappus croceus.*]
Arenaria arctica: **Lidia obtusiloba,** !ISC
Astragalus racemosus: **A. aboriginum** var. **glabriusculus,** !ISC
Cereus viridiflorus: **Echinocereus viridiflorus,** !ISC

Juniperus virginiana: **Sabina scopulorum,** !ISC
Draba alpina: **D. grayana,** !ISC [two printed labels, one for 1861 and one for 1862]
Erigeron acre: **Trimorpha elongata,** !ISC
Gentiana barbellata, isotype: **Gentianopsis barbellata** [The date, 1873, on the label, is incorrect. This was collected on Mount Flora, visited by Parry in 1861. A fine original watercolor, published as a line drawing, is affixed to the sheet.]
Helianthus petiolaris, !ISC
Penstemon caespitosus [Referred to on p. 254, "which Dr. Parry has detected the present season, and sent in a letter." This is abundant in Middle Park.]
P. cyathophorus, !ISC, "Middle Park" [This is common between Kremmling and Green Mountain Reservoir.]
Populus: **P. x acuminata,** !ISC
Saxifraga debilis: **S. rivularis,** !ISC
S. caespitosa: **Micranthes delicatula,** !ISC
S. integrifolia: **Micranthes oregana,** !ISC
Symphoricarpos occidentalis, !ISC [Not listed by Gray. Included in Hall & Harbour's list as no. 227.]
Townsendia grandiflora, !ISC

The following specimens, discussed by Engelmann in Parry and Engelmann (1862), were not in the Parry herbarium and presumably were unicates sent to Engelmann at the Missouri Botanical Garden (MO) at St. Louis.

Abies grandis: **A. bifolia,** !ISC, "Empire" [Clear Creek Co.]
A. douglasii: **Pseudotsuga menziesii**
A. menziesii: **Picea pungens,** isotype
A. nigra: **Picea engelmannii,** isotype
Pinus flexilis
P. ponderosa [subsp. **scopulorum**]
P. contorta var. **latifolia**

The Parry, Hall, and Harbour Expedition of 1862 3

Stimulated by his exploration of the Clear Creek Valley in 1861, Parry enlisted the help of two men from Illinois, Elihu Hall and J. P. Harbour, on a more ambitious expedition in the summer of 1862. Ewan and Ewan (1981) describe Hall (1820 [or 1822?]–1882) as a "botanist and farmer, born in Patrick Co., Va., one of the organizers of the Illinois Natural History Survey at Bloomington in 1858." Much less is known about Harbour (cf. Ewan and Ewan 1981, p. 95).

On this expedition, about 90 species of plants were discovered for the first time; thus, the specimens (referred to as "type specimens") that were used in publication of the descriptions and plates are invaluable. This is not to imply that these species are rare; indeed, most of them represent the most common and conspicuous species in the area, but the type specimens forever remain associated with the name and become the base reference for the description. However, several of the species (including *Ptilagrostis (Stipa) porteri* and *Trichophorum pumilum*) that the collectors found in South Park are very local, representing major disjunctions of species or genera recurring in the Northern Rocky Mountains, the American Arctic, or Central Asia. Some of these were not collected again until the the 1970s, and two of them, *Astragalus frigidus (americanus)* and *Campanula aparinoides,* have not yet been rediscovered in Colorado. At the present time, South Park's wetlands and fens, notably High Creek Fen (Mohlenbrock 1995), as well as the alpine slopes of the mountains to the west and north, appear to be the richest and most significant botanical refugia in the southern Rocky Mountains. Parry would have profited by a longer stay in the area, especially if he had visited Hoosier Pass, where *Eutrema edwardsii, Braya humilis, Saussurea weberi, Armeria scabra, Oxytropis podocarpa,* and other rare alpines occur, instead of the less interesting Georgia Pass.

Ewan and Ewan (1981) inaccurately represent the details of the 1862 expedition. For instance, they state (p. 169), "In 1862 in Colorado Parry met E. Hall and J. P. Harbour; perhaps they did some collecting together." Gray, as well as Parry, clearly stated in various publications relating to the 1862 expedition (one of which is presented in detail later on in this chapter) that the Parry, Hall, and Harbour party formed a team that set out together, remained together until shortly before the conclusion of the field work, and worked together in organizing the collections for identification and sale. Ewan and Ewan (1981, p. 94) also give an erroneous itinerary for the expedition, stating that "these plants were taken principally in the Clear Creek drainage, chiefly from Idaho Springs to Empire [City], across old Union Pass to Georgetown, and the present ski house at the foot of Loveland Pass, with side trips to canyons above Graymont (Bakerville) and Georgetown. They reached Mt. Breckenridge in Summit Co., probably via Hoosier Pass from South Park. Webster Pass was a trail in the 1860s and they may have traversed it." This statement better applies to the itinerary of the 1861 expedition. The detailed itinerary published by Engelmann (1868), given in this chapter, sets the record straight.

Parry's journal for the 1862 expedition is missing from the set in the Iowa State University Library archives. This is possibly the result of Parry sending his meteorological observations to Engelmann, who calculated the altitudes of points along the route by using data for St. Louis on the respective dates. Presumably the journal, if there was one, is still among the Engelmann papers at the Missouri Botanical Garden. Fortunately, the route taken by Parry, Hall, and Harbour in 1862 was published in Engelmann's paper, mainly dealing with the altitude of Pike's Peak (Engelmann 1868).

From Engelmann's paper, which lists altitudes in feet that he calculated from Parry's barometric readings at stated localities in chronological order and by stages, we can establish precisely the route taken by the collectors. The localities are listed in the following text, with explanations in modern terms in brackets as necessary.

ITINERARY

The following itinerary is of only very marginal value for ascertaining the exact localities where plants were collected, except for those species that are known to be rare and restricted to South Park, since the labels lack collection data. However, the route, in general, can be related to the modern road maps for the region. The party entered Colorado at Julesburg in the northeastern corner and essentially followed the South Platte River drainage to Denver. They then traveled west to Bergen Park and southward, along easy terrain, to near the present site of Conifer, where they followed the route of Highway 285, coming down Crow Hill to the North Fork of the Platte at Bailey. They then followed the Platte to Kenosha Pass, crossing into South Park and reaching Jefferson.

Mount Guyot from Michigan Creek, South Park, 1985. *Photo by David Hill.*

From Jefferson they diverged north along a wagon road to Georgia Pass and Mount Guyot, where they would have encountered the alpine flora for the first time. Returning to Jefferson, they followed the present route (Highway 77) southeast to Tarryall and Lake George. They then proceeded east along the present route of U.S. Highway 24 to Divide, Woodland Park, Manitou Springs, and Colorado Springs. At Colorado Springs, the party made a side trip to the summit of Pike's Peak. The party then returned to Denver along what is now the route of Interstate 25. Here Parry parted company with Hall & Harbour.

1. Route From Omaha to Denver	*Feet*
Omaha, library in statehouse, 211 feet above the Missouri	1,155
Baker's and Fales' Ranch, 2 miles above lower crossing of South Platte	3,161
Julesburg, upper crossing of South Platte, 8 feet above river	3,703
Mouth of Beaver Creek, 4 feet above river	4,284
Bijou Creek, on "cut-off"	4,712
Mail station, 12 miles from last point, now abandoned	4,963
Thirteen-mile Creek, 13 miles from Denver	5,776
Denver City, lower bottom of Platte, near mouth of Cherry Creek	5,303

2. Route From Denver Southwestward to Tarryall, 80 Miles

Mount Vernon, at the base of the mountains, 12 miles from Denver [creek in Jefferson County, a left-hand branch of Bear Creek, tributary to South Platte River]	6,421
Bergen's Ranch, 10 miles further southwest [Bergen Park, Jefferson County, southwest of Golden]	7,752
Bear Creek Station [creek in Jefferson and Arapahoe Counties, a left-hand branch of South Platte River]	7,198
Bradford Junction, 8 miles further	8,069
Dr. Casto's Ranch, 1Platte River]	8,380
Elk Creek, 6 miles from Junction	8,150
Summit of hill on road leading down to the Platte [Crow Hill]	8,881
North branch of South Platte	8,028
Same, 6 miles higher up, at Lee's Ranch	8,435
Same, 3 miles higher up, at a deserted ranch	8,657
Same, 11 miles higher up, junction of the upper branches, near a deserted ranch	9,153
First appearance of *Pinus aristata* on the road	9,342
Lake-house, on the Divide leading down to the South Park [Kenosha Pass; there is still a pond on the summit]	10,041
South Park, at Junction Ranch	9,453
Same, near the town of Jefferson	9,842
Range-house, near Georgia Pass	10,498
Georgia Pass, or Jefferson Pass, over the Snowy Range to Georgia Gulch [There is a good road from Jefferson north to the pass.]	11,487
Mount Guyot, west of the pass	13,223
Tarryall, a deserted town [29 mi. east and 6 mi. south of Fairplay, 39°06′5″N, 105°28′21″W, T11S R72W S5, below Tarryall Reservoir, Tarryall Quadrangle]	9,932

3. Route From Tarryall Eastward to Colorado City [Colorado Springs], 70 Miles

This route would be the present Jefferson–Lake George–Manitou–Colorado Springs road.

Eastern edge of South Park, on Tarryall or Middle Branch of the South Platte	9,538
On road to Colorado City [which antedated Colorado Springs], 20 miles east	8,895

At the foot of a high hill, where the road leaves the Middle Branch	8,503
On South Branch of the South Platte, where the road crosses it at Gleason's Ranch, 46 miles from Tarryall	8,151
Near the Divide between the waters of the Platte and the Arkansas, about 10 miles west of Pike's Peak [This would be in the vicinity of Lake George.]	8,724
Near the western base of the peak [probably in the vicinity of Divide]	9,327
Upper course of Fontaine-qui-bouit [Fountain Creek], where the road first strikes it, north of the peak [Woodland Park]	8,273
On Fontaine-qui-bouit, 3 miles further down	7,794
Soda Springs [Manitou Springs], at the eastern base of the peak	6,515
Last timber growth on the north slope of the peak [= timberline] 12,043	
Summit of Pike's Peak	14,216
Colorado City, on the Fontaine-qui-bouit	6,342

4. Route From Colorado City Northward to Denver, 70 Miles

On road, 10 miles north of Colorado City	6,753
Garlick's Ranch, toward the Divide	7,105
Divide of Arkansas and Platte Rivers [vicinity of Greenland]	7,554
Plum Creek, near its head, 8 miles north of last station [between Larkspur and Castle Rock]	6,840
Same, 30 miles south of Denver	6,409
Denver City	5,303

5. Route From Denver Westward to Empire City, 50 Miles

From here on Parry was traveling alone.

Mount Vernon, as above	6,421
Junction of north and south fork of Clear Creek [the present road junction, the north fork going to Blackhawk and the south fork to Idaho Springs]	7,086
Idaho [now Idaho Springs]	7,800
Head of Virginia Gulch, a high divide between North and South Clear Creek [Virginia Gulch is the old stage road from Idaho Springs to Central City.]	9,690
Missouri City (Consolidated Ditch Office), near Central City	9,073

[Parry evidently then returned to Idaho Springs.]

South Clear Creek, at mouth of Fall Creek, 3 miles above Idaho [present junction with Fall River road to Mary's Lake]	7,930
Level of Clear Creek, at Empire City	8,583

6. *Route From Empire City Northwestward to Hot Springs, in Middle Park, 50 Miles*

Lindstrom's Mill, on Mad Creek, 1 mile above Empire City [This is along the present road to Berthoud Pass.]	8,738
Survey Station No. 50, 1	
Deserted ranch, 5 miles above Empire City, at base of Berthoud's Pass	9,464
Little Park, a survey station below the Pass (*Primula parryi* abundant)	10,715
Summit of Berthoud Pass, with large timber	11,349
Brush shanty, a survey station, 1 mile from the pass, on its western slope	10,696
Limit of trees on the range west of the Pass	11,816
Head of Middle Park, where the first open ground begins	8,690
Hot Springs [now Hot Sulphur Springs] of Grand River, 25 miles below last station	7,725

7. *Route From Empire City Southwestward to Gray's Peak, 20 Miles*

Empire City, as above	8,583
Georgetown, 6 miles further south	8,452
Limit of trees on eastern slope of ridge leading to Gray's Peak	11,643
Summit of Gray's Peak	14,251

8. *Towns in Colorado Territory*

Denver City	5,303
Colorado City	6,342
Mount Vernon	6,421
Georgetown	8,452
Empire City	8,583
Missouri City	9,072
Jefferson, in South Park	9,842
Tarryall (a deserted town)	9,932

9. *Passes*

Georgia Pass, from South Park to Middle Park	11,487
Berthoud's Pass, from the Clear Creek Valley to the head of Middle Park	11,349

10. *Limit of Trees*

On the Snowy [Front] Range, eastern slope of Gray's Peak	11,643
On the south slope of Mount Flora [see the list that deals with the next leg of the trip]	11,807
On the range west of Berthoud's Pass, north slope	11,816
On the northern slope of Pike's Peak	12,043

11. *Alpine Summits*

Mount Flora, a detached peak, east of Parry's Peak [Mount Flora, named by Parry, is about 2 mi. east of Berthoud Pass, on a spur ridge leading northeast 2 mi. to Parry's Peak, and connecting to the Continental Divide a mile or so farther with James Peak.]	12,878
Parry's Peak (so named by Surv. Gen. F. M. Case), a peak of the Snowy Range, northwest of Empire City	13,133
Mount Guyot, near Georgia Pass	13,233
Pike's Peak	14,216
Gray's Peak, southwest of Empire City	14,245
Mount Engelmann, 2 miles southeast of Berthoud Falls	13,362

Ascent of Pike's Peak

An interesting account of the ascent of Pike's Peak on July 1, 1862, was published in Trans. Acad. Sci. St. Louis 2(1):120–133.1863, from a letter addressed to Torrey and communicated by him.

Dear Sir: In accordance with frequent suggestions from you, recommending the examination of the memorable botanical locality known as James', or Pike's Peak, I feel gratified in being able to furnish you with a brief sketch of the results of such an exploration, accomplished on the 1st of July, 1862.

Since Dr. Edwin James, of Col. Long's expedition, first visited this alpine summit forty-two years ago, on the 14th of July, 1820, there is no record of any professed botanist having made the ascent. For this long period, its peculiar vegetation has bloomed unheeded, and the meagre collection of plants made by Dr. James has not been duplicated in scientific herbaria.

It is true indeed, that of late years, since the rapid settlement of the adjoining region, popularly known as Pike's Peak, various pleasure parties, intent on sight-seeing, and even *ladies,* have ventured to the snow-crowned summit, and Mr. M. S. Beach, of Colorado City, our guide on this last occasion, counted it as his third ascent; but by all these its floral treasures were only casually observed, and in no instance that I can learn have botanical collections been made.

The truthful and graphic account given by Dr. James, in Long's Expedition, of the ascent of this *"highest peak,"* shows that the route then taken was substantially the same as that followed by us, and is no doubt the one most accessible, at least from the northern slope.

That remarkable and interesting stream, known by the expressive French name of *Fontaine-qui-bouit,* which circles round the gigantic mass of rocks comprising the main peak, together with its lower range of mountains, pursues a general course east of south, and collects the waters flowing from its northern and eastern slope. This drainage is effected through numerous tributaries, coming more or less direct from the main peak, and cleaving their way through chasmed valleys and *cañons* of the most rugged character. Up one of the main forks, which enters the principal stream at the noted locality called Soda Springs, or Boiling Fountain, lies the most direct route for making the ascent. This stream, which ought to receive the historical name of *James Creek,* at its junction with the *Fontaine-qui-bouit,* is about six feet in width, but soon contracts its dimensions, as the valley through which it descends becomes narrow; and, farther up, is obstructed by fallen rocks.

Winding among these, its swift current rushes along till a sudden descent projects it in the form of rapids and falls. Here the scenery is wild in the extreme; the torrent, often lost entirely to view among masses of rock, shoots forth in frequent cascades, or is seen, through the vista of overhanging cliffs, boiling along its tortuous channels. The trail here is a succession of tedious clamberings from one mass of rock to another, or winding along the steep verge of precipices, and over sloping banks of decomposing granite. Among these rock crevices grows profusely the elegant flowering shrub *Rubus* [= *Oreobatus*] *deliciosus* of

James, now just passed out of flower, and maturing its reddish purple fruit. This latter, however, will hardly be found to merit the title of *delicious,* the mass of the berry being composed of large grains, with very meagre insipid pulp. An interesting associate of this common shrub is the *Jamesia americana* Torrey & Gray, its neat, white flowers contrasting prettily with its wrinkled, velvety leaves. Conspicuous among other plants may also be noted the *Yucca angustifolia* [= *Y. glauca*], now in the full glory of its globe-shaped flowers, of satiny lustre. Here and there, also, in detached localities, *Penstemon torreyi* [= *P. barbatus* var. *torreyi*] Bentham, sends up its brilliant red spikes.

The trees include the elegant pyramidal forms of *Abies grandis* [= *A. concolor*], here remarkable for the unusual length and breadth of its leaves; *Abies douglasii* [= *Pseudotsuga menziesii*] is also common, associated with *Abies menziesii* [= *Picea pungens*], and *Pinus ponderosa.* Succeeding these at a higher elevation comes the Pine, which must now be regarded as the original *Pinus flexilis* James. As such it has for several years been recognized by Dr. Engelmann, from the collection of Fendler and others, though still doubtfully regarded in late European works on Coniferae. Dr. James' account of this tree being quite meagre and in some respects contradictory, it may be satisfactory to dwell at some length on its peculiar habit, as exhibited in this its original locality. In general appearance it very closely resembles our *P. strobus,* from which it differs mainly in its shorter and stouter entire leaves, more branching mode of growth, as well as in the yellowish brown cones with peculiar ligneous scales. The cones are inclined to be pendulous. The fertile aments occupy the extremity of the growing branch, extending in the same line with it; but in the second year the terminal bud shoots out, and by its development the growing cones one to five together, are gradually deflected. Rarely more than two of these become fully grown, and as a general rule the mature cones fall off at the close of the second year, the opening scales having previously dropped their winged seeds. These seeds are nearly equal in size to those of the New Mexican nut-pine, *Pinus edulis,* of an irregular oval form, 4–5 lines long [a line being 1⁄12 inch], and possess similar edible qualities. In addition to other peculiarities of this pine may be noticed its slowness of growth; thus, on a small trunk of 7¾ inches in diameter, there were 232 annual rings. Its wood is soft, of fine texture, the heart wood inclined to a yellowish caste. The flexibility of its branches, on which Dr. James founded its specific name, is partly due to the thickness of the elastic bark of the smaller twigs. The bark of the trunk is of a dark reddish-gray color, considerably furrowed, and about equal in thickness to that of our common white pine. The average height of full-grown trees is from 40 to 50 feet; they have a rounded outline, are generally low branched, and spreading; in the largest specimens observed, the trunk, a short distance from the surface of the ground, had a diameter of two feet and upwards.

The vertical range of this species, as observed between latitude 38° and 40°N, is from 7,000 to 11,000 feet above the sea. [Actually a lower value of 5,000 feet has been observed at the Pawnee Buttes in Weld County.] It rarely occurs in large bodies of timber, but is mostly of scattered growth, being associated, at its lowest range, with *Pinus ponderosa* and *Pinus contorta*, and at its upper limits with *Pinus aristata* and *Abies* [= *Picea*] *engelmannii.* Besides *Pinus flexilis,* which alone seems to have particularly attracted the attention of Dr. James, he mentions, in a cursory way, the occurrence of *Abies balsamea, A. canadensis, A. alba, A. nigra,* and *A. rubra,* these being the then recognized representatives of the fir tribe in eastern North America. In this enumeration, the very common error of confounding analogous species was committed, an error to which those who simply observe, and do not collect, specimens, are quite apt to fall into. It is sufficient to state, in this connection, that not a single one of these species is recognized at present as occurring in this part of the Rocky Mountains; in fact, most of the species there met with were, at that early day, unknown to science. Under the names of *Abies nigra* and *A. rubra,* there is little doubt that Dr. James had in view a very puzzling Rocky Mountain species which, in imperfect material, has frequently turned up in collections from this region, as being usually classed under the names of *A. alba* or *A. nigra.* My attention having been particularly directed to this species by Dr. Engelmann, I became soon satisfied, in pursuing the investigation, that this was in fact a single undescribed species, appearing under different forms according to soil, altitude, and exposure, to which, accordingly, I have ventured to affix the name of its actual discoverer, calling it *Abies* [= *Picea*] *engelmannii.*

I continue the narrative of our ascent. On reaching an elevation of about 9,000 feet, the contracted valley up which we were travelling spread out into more free stretches, being on a level with the plateau of the first range of foothills. The surface is here covered with a rank growth of grass and scattered pine timber. Sweeping fires, which had passed over nearly this entire region of country, occasioned the destruction of the principal pine growth which, with its dry, naked trunks, gave a somewhat forbidding aspect to the more open scenery. In the moist, lower valleys the fallen timber unites with a matted growth of subalpine willows, rendering the passage tedious and difficult. The several valleys here converging from different directions gradually merge into a steeper, rocky slope, occupied, as before, with dead wood. On this subalpine inclination, few peculiar plants are met with; *Penstemon glaucus* [= *P. whippleanus*] and *P. alpinus* [= *P. glaber*] being most conspicuous. From this point, the mountain slope increases quite rapidly, and the ascent is by steady and continuous climbing. The timber growth, confined almost exclusively to a more stunted form of *Abies engelmannii,* with scattering trees of *Pinus flexilis,* soon gives place to open patches, disclosing a vegetation purely alpine. Here, for the first time, *Pinus aristata* Engelm. makes its appearance; its deformed trunks, beset with withered branches and sending off leafy tufts close to the ground, serve to give a peculiar blighted look to the landscape which it occupies. I have had frequent occasion, in my various mountain rambles, to notice the abruptness with which the alpine flora usually makes its appearance. After toiling slowly up the steep ascent, with little or nothing new to attract the attention, suddenly, on mounting some exposed knoll, a profusion of alpine flowers bursts on the view. The plants thus met with include, almost constantly, *Primula angustifolia, Cymopterus alpinus* [= *Oreoxis humilis*], *Eritrichium* [= *Eritrichum*] *aretioides, Arenaria arctica* [= *Lidia obtusiloba*], *Silene acaulis, Aplopappus* [= *Tonestus*] *pygmaeus,* etc. From this point, there is a constant succession of these interesting forms, varied according to the peculiar exposure and the character of rock or soil. Along the somewhat scanty alpine brooks of this region, I was pleased to notice the elegant flowered *Primula parryi,* with its very constant associate, *Sedum* [= *Clementsia*] *rhodanthum.* The *Mertensia sibirica* [incorrect reference to *M. ciliata*] still maintains its position by the edges of streams, extending thence downward to the very base of the mountains. Here also we more or less constantly meet with *Sibbaldia procumbens, Saxifraga cernua, S. debilis* [= *S. hyperborea* subsp.], *S. punctata* [incorrect reference to *Micranthes odontoloma*], *Caltha* [= *Psychrophila*] *leptosepala,* and others. Among the plants not heretofore observed is the neat *Androsace chamaejasme,* which exhales a pleasant odor of bitter almonds, and the beautiful red-flowered *Saxifraga* [= *Telesonix jamesii*], rooting in crevices on the vertical walls of shaded rocks. These various forms continue, intermixed with patches of snow, till the limit of arborescent growth is reached, observing a well defined horizontal line along the mountain slope. According to barometric measurement this line, at the point observed, having a northeast exposure, shows an elevation of 12,043 feet above the sea level. The last trees to maintain their position in this exposed locality are *Abies engelmannii* and *Pinus aristata,* both of them dwarfed and stunted in their struggle with the elements and exhibiting marks of decrepit age, in blasted trunks and prostrate branches. From some of these alpine *centenarians* we made huge fires to keep off the chilly night air, while spruce boughs supplied us with spring mattresses.

As the setting sun passed over the western slope, the gigantic outline of Pike's Peak was projected on the plain below with wonderful distinctness and in massive proportions.

Astir by daylight to watch, from our mountain eyrie, the glories of an unclouded dawn, we were surprised and gratified by the faint chirp of birds, strangely contrasting with the bleak scenery by which we were surrounded. This morning carol we afterwards found to proceed from a species of mountain swallow [perhaps a pipit?], the nest of which we discovered still higher up on the mountain slope at an elevation of not less than 13,000 feet above the sea. We could not but admire the taste with which the selection was made; a snug recess, scooped out amid the matted foliage of *Silene acaulis,* concealed from view by an overhanging tuft of *Dryas octopetala,* crowded with its pure white blossoms; while, in close vicinity, bloomed the beautiful *Primula angustifolia* and fragrant *Eritrichium* [*Eritrichum*] *aretioides.* Under such circumstances, natural feelings overcame the scientific taste for collecting, and we left undisturbed the nest with its contents, consisting of five mottled, granite-colored eggs.

As the sun rose majestically above the well-defined horizon of the plains, the resemblance to a wide open sea was strikingly manifested. A slight haze served to heighten the pleasant illusion, the inconsiderable elevations appearing only as ripples or low islands on its surface. To carry out the resemblance still farther, the rounded grassy swells and reeflike ledges of tilted rock at the foot of the mountains could readily be taken for surges and breakers on this once well-defined coast.

Setting our faces once more towards the gigantic peak, still towering 2,000 feet above us, we commenced the final ascent, slowly mounting over a varied surface composed of disintegrating rock interspersed with patches of alpine sward. Conspicuous among the plants decorating this mountain sod were the bright azure flowers of *Mertensia paniculata* [incorrect reference to *M. alpina*] and *Eritrichium aretioides,* the latter, as one of the party significantly suggested, resembling "a piece of the sky just fallen down." Though as late in the season as the first of July, all the indications of vegetation were those of early spring. I looked in vain at the foot of the snowdrifts to find the *Chionophila* (snow-lover), discovered here by Dr. James, and found last year on the Snowy Range, but the season was, no doubt, too early; *Trifolium nanum* and *T. dasyphyllum* were, however, in full bloom and quite conspicuous. Near the very summit we first came upon the interesting taprooted *Claytonia* [*megarhiza*], observed so abundantly last season on the Snowy Range at the head of South Clear Creek. Here it seemed dwarfed and stunted, having far less conspicuous leaves and flowers. This plant, together with an alpine *Thlaspi* [= *Noccaea montana*], were the only flowers in bloom on the highest elevation.

The summit gained, there was opened an extensive view towards all points of the compass. To the east stretched the unlimited expanse of the great plains, while to the south could be traced the course of the upper Arkansas; north and west was a confused mass of mountains interspersed with open valleys, including the broad basin of South Park, bounded by the sharply defined outline of the Snowy Range. From this point I was able to detect an elevated peak in the Snowy Range, visited a few weeks previously, having an elevation, according to barometric measurement, of 13,223 feet above the sea. I have called this peak *Mount Guyot,* in compliment to the distinguished Swiss-American savant of that name. Other still more elevated points could be noticed, some of which are perhaps as high or even higher than Pike's Peak.

The summit of Pike's Peak is a somewhat level plateau embracing several acres in extent, strewn with masses of detached rocks of a fine-grained granite, and occupied in part by extensive snowdrifts. On the highest point of one of these, by the aid of a rough tripod made from climbing staffs brought up by the ascending party, I set up my barometer which, on adjusting the column of mercury, stood at 18.100; attached thermometer, 45°F.; detached thermometer, 37°F. Chilly gusts of wind, sweeping over the bald exposure, compelled me to change the place of observation to a more sheltered spot about fifteen feet below the main summit. At this point I made a series of observations for ascertaining the elevation, giving a result, as computed by Dr. Engelmann, of 14,216 feet above the sea.

Our observations finished, the more *facilis decensus* was commenced, not, however without many weary steps and much carefully poised balancing. We reached the timber line to partake of our last mountain meal, and thence, by nightfall, our pleasant campground at the *Fontaine-qui-bouit.*

List of Plants

The labels accompanying the collection have one of three headings: *American Plains Flora, Lat. 41°; Rocky Mountain Alpine Flora, Lat. 39–40°;* or *Rocky Mountain Flora, Lat. 39–41°* (Fig. 3). A curious fact about the DUKE collection is that a large number of the specimens were without any labels, bearing only the Hall & Harbour numbers. These would be of very little value save for the existence of the published list by Gray. One wonders if Hall possibly ran out of labels and sent out some sets or parts of sets without them.

The Hall & Harbour collections have virtually nothing in the way of locality data, and they lack dates or altitudes. One would think they might be rather worthless scientifically. Even the type specimens lack any information that would be helpful in determining the exact localities. Nevertheless, their collections represent a surprisingly complete coverage of that area of the Rocky Mountains and comprise the finest assembled up to that time. The scientific knowledge of the flora, it appears, must have been provided by Parry. The specimens were numbered (with few errors of families) according to the taxonomic arrangement of the day and were not dated, making it impossible to relate the collections to routes of travel. There are, however, a few clues to be found in the kinds of labels used. The 1861 Parry collections indicated the collector and the year. Many of the 1862 Parry labels also gave this much information, but evidently these labels were exhausted, and many of them (often with numbers cited by Gray as Hall & Harbour's) have the bottom line (collector and year) cut off with a scissors. Despite the lack of locality data, the headings on the printed labels give a rough idea of the general area of collection. This was not an uncommon practice worldwide. A very small number of specimens were evidently collected in western Nebraska or Kansas.

Among the two sets, there are numerous disagreements as to the species present. Some numbers, here marked "no entry," were not seen by Gray; some of these turn up in the ISC or DUKE set. Such confusions and omissions suggest a rather hasty division of the duplicates into sets. Even Parry's personal set is not complete.

The following enumeration of the specimens uses the sequence of the numbers written by the collectors on their labels and followed by Asa Gray in his account (Gray 1863). Although the numbers on the Hall & Harbour labels are generally reliable, Parry seems to have added his own, usually irrelevant labels, often bearing the date 1861. Specimens not seen in either the DUKE or ISC collections are given the equivalent current name, an educated guess on my part. It must be borne in mind that errors did occur

Rocky Mountain Flora, Lat. 39°-41°.
No. 409
E. HALL & J. P. HARBOUR, Colls, 1862.

ROCY MOUNTAIN ALPINE FLORA, LAT. 39°-41°.
No. 391 Pentstemon confertus Dougl
E. HALL & J. P. HARBOUR, Colls, 1862.

Rocky Mountain Flora, Lat. 39°-41°.
No. 56 Viola canina L. var adunca
E. HALL & J. P. HARBOUR, Colls. 1862.

AMERICAN PLAINS FLORA, LAT. 41°.
No. 390 Pentstemon acuminatus Dougl
E. HALL & J. P. HARBOUR, Colls, 1862.

No. ROCKY MOUNTAIN FLORA.
COLORADO TERRITORY, lat. 39° - 41°, Alpine and Subalpine.
Rubus —
C. C. PARRY, Coll. 1862.

No. 75 — ROCKY MOUNTAIN FLORA.
COLORADO TERRITORY, lat. 39° - 41°.
Thalictrum alpinum
C. C. PARRY, Coll. 1862.

Six representative labels of the Parry, Hall, and Harbour 1862 expedition.

in the original distribution, and it is very likely that in some instances a specimen at Gray Herbarium may not match its equivalent in the DUKE or ISC collection.

Ewan and Ewan (1981) provide an interesting insight into the disposal of the collections. They quote (p. 94) a letter from Leo Lesquereux to H. N. Bolander: "I have succeeded in selling 8 sets of Hall to M. Boissier. They are cheap. But Mr. Hall wants the money immediately for building a house and of course must make a heavy discount to sell them all at once."

The Hall & Harbour Colorado collections of 1862 were the largest and most important ones gathered between the time of Edwin James and the twentieth century. Parry, Hall, and Harbour must have worked very diligently to assemble what became an extraordinary collection, the largest yet made in Colorado in a single season, of almost 700 numbers, replicated in perhaps as many as ten or more duplicate sets (7,000 specimens!). Asa Gray identified one set of these and, as an introduction, it is useful to quote some of the opening remarks of Gray in his enumeration (Gray 1863):

> An interesting account by Dr. Parry of his first explorations of the Rocky Mountains in Colorado Territory, made in the summer of 1861, was published in the American Journal of Science and Arts, vol. 33, 1862. This was followed by an enumeration of the plants in the choice botanical collection which he made, as determined by myself, Dr. Engelmann, and others. The importance of this pioneer exploration, both in a physico-geographical and a botanical point of view, decided Dr. Parry to repeat and extend it the following year, to undertake more full and exact observations upon the configuration of the district, and the altitude of the loftier peaks, and to secure a larger botanical collection. In the latter view, Dr. Parry was joined by two zealous and enterprising botanical companions, Messrs. Hall and Harbour, of Illinois, who devoted their entire energies to the collection of plants. The botanical collection, accordingly, through these conjoint labors and explorations, is full, excellent, and of great interest. . . .
>
> It should be remarked that the general collection, although made by the three associates conjointly, is distributed under the tickets of Messrs. Hall and Harbour,—upon whom indeed the labor of the collection more immediately devolved—and is numbered quite independently of Dr. Parry's collection of 1861, thus avoiding all danger of confusion between the two. But a small separate collection made by Dr. Parry late in the summer, at stations visited by himself alone, which supplements or helps out the general collection, bears Dr. Parry's numbers

of the former year . . . or, when of plants not in that collection, the numbers are in continuation of it. . . .

The plants were numbered and distributed into sets by Messrs. Hall and Harbour before they were seen by me, and a full set was supplied to me for examination.

Dupree, based on letters exchanged between Gray, Parry, and Hall & Harbour, writes (1959, p. 325): "Gray performed his usual service of sorting the collections and publishing accounts of them as they came in. Parry had collected alone in 1861, and with Elihu Hall and J. P. Harbour in 1862. The team had the kind of dispute so common among both old and new field botanists. Hall and Harbour made many sets of each species, while Parry took pains with a smaller number, opening himself to the charge of being lazy. Gray was the one who heard their case and decided on the disposition of the joint collection, apparently to the satisfaction of all concerned."

1. *Atragene alpina:* **A. columbiana,** !DUKE, ISC
2. *Clematis douglasii:* **Coriflora hirsutissima,** !DUKE
3. **C. ligusticifolia,** !DUKE, ISC
4. *Pulsatilla nuttalliana:* **P. patens** subsp. **multifida,** !DUKE, ISC
5. **Anemone multifida** var. **globosa,** !DUKE [with a contaminant stem of no. 6, *Anemone caroliniana*], ISC
6. **A. caroliniana,** !DUKE, ISC [This must have been collected in Kansas.]
7. *A. narcissiflora:* **Anemonastrum narcissiflorum** subsp. **zephyrum,** !DUKE, ISC
8. **Thalictrum fendleri,** !DUKE, ISC [cited by Boivin, Rhodora 46:442.1944, as *T. confine* var. *greeneanum*]
9. **T. sparsiflorum,** !DUKE, ISC [labeled *T. fendleri*]
10. **T. alpinum,** !DUKE, ISC, isotype of *T. scopulorum*
11. *Ranunculus cymbalaria:* **Halerpestes cymbalaria** subsp. **saximontana,** !DUKE, ISC
12. *R. hyperboreus* var. *natans:* **R. hyperboreus** subsp. **intertextus,** !DUKE, ISC
13. *R. nuttallii:* **Cyrtorhyncha ranunculina,** !DUKE, ISC [as *Cyrtorhyncha!*]
14. **R. eschscholtzii,** !DUKE [mounted with no. 14], ISC
15. *R. affinis* var. *leiocarpus:* **R. pedatifidus,** !DUKE, ISC
16. *R. affinis* var. *cardiophyllus:* **R. cardiophyllus,** !DUKE [with *R. pedatifidus* and *Thalictrum alpinum*], ISC

17. **R. adoneus,** !DUKE, ISC
18. *R. flammula* var. *reptans:* **R. reptans** var. **ovalis,** !DUKE, ISC
19. **R. alismifolius,** !ISC
20. **Myosurus minimus,** !DUKE, ISC
21. *Caltha leptosepala:* **Psychrophila leptosepala,** !DUKE, ISC
22. *Trollius laxus* var. *albiflorus:* **T. albiflorus,** !DUKE, ISC
23. *Aquilegia vulgaris* var. *brevistyla:* **A. saximontana,** !DUKE, ISC
24. **A. coerulea,** !DUKE, ISC
25. *Delphinium elatum:* **D. barbeyi,** !DUKE, ISC
26. *D. scopulorum:* **D. x occidentale,** !DUKE, ISC
27. *D. scopulorum* (high alpine form)*:* **D. ramosum** var. **alpestre,** !DUKE, ISC [This is known from Boreas Pass, near Georgia Pass, where collections were made.]
28. *D. menziesii:* **D. nuttallianum,** !DUKE, ISC
29. *Aconitum nasutum:* **A. columbianum,** !DUKE, ISC
30. *Berberis aquifolium:* **Mahonia repens,** !DUKE, ISC
31. *Corydalis aurea* var. *curvisiliqua:* **C. aurea,** typical, !DUKE, ISC [DUKE specimen has two plants of *Cardamine pensylvanicum* mounted on the same sheet, no. 33.]
32. *Nasturtium obtusum:* **Rorippa teres,** !DUKE, ISC
33. *Cardamine hirsuta:* **C. pensylvanica,** !DUKE, ISC
34. **C. cordifolia,** !DUKE, ISC [Gray correctly noted a relationship to the Alpine European *C. asarifolia.* The eastern Siberian species, *C. pedata,* is also a likely close relative. Few species of the genus have simple leaves.]
35. *Streptanthus angustifolius:* **Boechera drummondii,** !DUKE, ISC
36. *Turritis patula:* **Boechera cf. Retrofracta** [No. 36, !DUKE, ISC, is labelled *Arabis holboellii* and is a mixture of **Boechera fendleri,** and **B. lignifera.** On this sheet is a scrap label, "*Turritis patula* Graham, Upper Clear Cr., Parry, 94," but this label does not appear to apply to the specimens.]
37. *Sisymbrium virgatum:* **Halimolobus virgata,** !DUKE, ISC [known in Colorado only from South Park]
38. **Erysimum cheiranthoides** subsp. **altum,** !DUKE, ISC
39. *E. asperum* var. *pumilum:* **E. capitatum,** !DUKE
40. *Sisymbrium sophia* [*canescens* on the label]*:* **Descurainia incana,** !DUKE, ISC. [Gray's notes imply that two species are involved here, probably *Descurainia pinnata* and *D. incana.*]
41. **Draba crassifolia,** !DUKE, ISC (specimen a) [with *Parry 93, 95,* Gray's Peak. Gray noted that his specimen of this number contained a mixture of three species. The DUKE sheet also contains b, **D. albertina,** and c, **Erysimum asperum.**]

42. **Draba nemorosa,** mixture, **D. nemorosa** and **D. albertina,** !DUKE, ISC [*D. nemorosa* only]
43. **Smelowskia calycina,** !DUKE, ISC [This was not in Gray's list.]
44. **Draba aurea,** !ISC [with *Parry* (1861) *103*], DUKE
45. **D. streptocarpa,** !DUKE
46. *Thlaspi cochleariforme:* **Noccaea montana,** !DUKE, ISC
47. *Physaria didymocarpa:* **P. vitulifera,** !ISC [with *Parry 101*], DUKE
48. *Vesicaria ludoviciana:* **Lesquerella ludoviciana,** !ISC [with *Parry* (1862) *114*], DUKE
49. *V. montana:* **Lesquerella montana,** isotype of *Vesicaria montana,* !DUKE
50. *Stanleya integrifolia:* **S. pinnata** var. **integrifolia,** !DUKE, ISC [The type collection of James came from near Colorado Springs. Both leaf forms are represented on the specimens of the ISC sheet.]
51. **Thelypodium integrifolium,** !DUKE, ISC
52. *Cleome integrifolia:* **C. serrulata,** !DUKE, ISC
53. *Cleomella tenuifolia:* **C. angustifolia,** !DUKE, ISC
54. **Viola biflora,** !DUKE, ISC
55. **V. nuttallii,** !DUKE, ISC
56. *V. muhlenbergii* var. *pubescens:* **V. labradorica,** !DUKE, ISC
57. *Ionidium lineare:* **Hybanthus verticillatus,** !DUKE, ISC
58. *Hypericum scouleri:* **H. formosum,** !DUKE, ISC
59. *Elatine americana:* **E. triandra,** !DUKE, ISC
60. **Glaux maritima** [var. **angustifolia**], !DUKE
61. **Silene scouleri,** [subsp. **hallii**], isotype of *S. hallii,* !DUKE, ISC
62. *S. drummondii:* **Gastrolychnis drummondii,** !DUKE [with a stem of no. 64, *Anotites menziesii*], ISC
63. *Lychnis apetala:* **Gastrolychnis kingii,** syntype of *Lychnis montana,* !DUKE, ISC
64. *Silene menziesii:* **Anotites menziesii,** !DUKE, ISC
65. **S. acaulis** L. subsp. **subacaulescens,** !DUKE, ISC
66. **Paronychia pulvinata,** isotype, !DUKE, ISC
67. **P. jamesii,** !DUKE, ISC
68. *Sagina linnaei:* **S. saginoides,** !DUKE, ISC
69. *Arenaria rossii:* **Alsinanthe stricta,** !DUKE, ISC [same, mixed with **Tryphane rubella,** !DUKE]
70. **Stellaria umbellata,** !ISC
71. **S. longipes,** !ISC
72. *S. borealis:* **S. calycantha,** DUKE, ISC
73. **S. umbellata,** !DUKE, ISC
74. **Moehringia lateriflora,** !DUKE, ISC

75. *Cerastium arvense:* **C. strictum,** !DUKE, ISC [with the DUKE specimen a stem of *S. umbellata*], plus *C. vulgatum?* var. *beeringianum:* **C. beeringianum**
76. **Stellaria longipes,** !DUKE, ISC
77. *Arenaria arctica:* **Lidia obtusiloba,** !DUKE, ISC
78. *Stellaria jamesii:* **Pseudostellaria jamesiana,** !DUKE, ISC
79. *Arenaria fendleri:* **Eremogone fendleri,** !DUKE, ISC
80. **Limosella aquatica,** !DUKE, ISC
81. **Talinum parviflorum,** !DUKE, ISC
82. *Claytonia virginica:* **C. lanceolata,** !DUKE, ISC
83. *C. arctica* var. *megarhiza:* **C. megarhiza,** !DUKE
84. *C. chamissonis:* **Crunocallis chamissoi,** !DUKE
85. **Sidalcea candida,** !DUKE
86. *Malvastrum coccineum:* **Sphaeralcea coccinea,** !DUKE
87. *Linum perenne:* **Adenolinum lewisii,** !DUKE, ISC
88. **Geranium richardsonii,** !DUKE, ISC
89. *G. fremontii* var. *parryi:* **G. caespitosum,** !DUKE, ISC
90. **Ceanothus fendleri,** !DUKE, ISC
91. *C. ovatus:* **C. herbaceus,** !DUKE, ISC
92. **Pachy[xi]stima myrsinites,** !DUKE
93. **Acer glabrum,** !DUKE
94. **Lupinus pusillus,** !DUKE
95. *L. ornatus:* **L. plattensis,** isotype of *L. ornatus* var. *glabratus,* !DUKE, ISC
96. *L. caespitosus:* **L. lepidus** subsp. **caespitosus,** !DUKE [with a stem of *Trifolium parryi*]
97. **Trifolium dasyphyllum,** !DUKE
98. **T. parryi,** !DUKE, ISC
99. **T. nanum,** !DUKE
100. *Dalea laxiflora:* **D. enneandra,** !DUKE [collected on the eastern plains]
101. *Psoralea lanceolata:* **Psoralidium tenuiflorum,** !DUKE
102. *P. floribunda:* **Psoralidium tenuiflorum,** !DUKE, ISC
103. *P. argophylla:* **Psoralidium argophyllum,** !DUKE, ISC
104. *Dalea alopecuroides:* **D. leporina,** !DUKE
105. *Petalostemon macrostachyum:* **Dalea cylindriceps,** !DUKE, ISC
106. **Astragalus kentrophyta,** [subsp. **implexus**], !DUKE
107. *Thermopsis rhombifolia* and *T. montana:* **T. divaricarpa,** !DUKE; **T. rhombifolia,** !DUKE, ISC
108. *Hosackia purshiana:* **Lotus wrightii,** !DUKE, ISC [collected east of Colorado on the plains]

109. *Lathyrus ornatus* and a pubescent variety: **L. eucosmus,** !DUKE, ISC
110. *L. linearis:* **Vicia americana** var. **minor,** !DUKE, ISC [also 109b in DUKE collection]
111. *L. polymorphus:* **L. polymorphus** subsp. **incanus,** !DUKE, ISC
112. *L. palustris* var. *myrtifolius?:* **L. leucanthus,** !DUKE, ISC
113. **Astragalus racemosus,** !DUKE, ISC
114. **A. bisulcatus,** !DUKE [a detached fruiting branch belongs to **A. aboriginum**]
115. *A. nigrescens:* **A. tenellus,** !DUKE, ISC
116. *A. glabriusculus* var. *major,* isotype*:* **A. aboriginum,** !DUKE, ISC [annotated by Thomas N. Kaye as *A. australis* var. *glabriusculus* (Gray) Isely]
117. *A. oroboides:* **A. eucosmus,** !DUKE, ISC
118. **A. flexuosus,** !DUKE
119. **A. gracilis,** !DUKE, ISC
120. **Oxytropis deflexa** var. **sericea,** !DUKE, ISC
121. *Astragalus cf. debilis:* isotype of **A. hallii,** !DUKE, ISC
122. **A. mollissimus,** !DUKE [One specimen is **Oxytropis sericea.**]
123. **A. parryi,** !DUKE, ISC
124. **A. drummondii,** !DUKE (with stem of *A. bisulcatus*)
125. **A. alpinus,** !DUKE
126. *A. cyaneus:* **A. shortianus,** !DUKE, ISC
127. **A. missouriensis,** !DUKE
128. **A. sparsiflorus,** isotype, !DUKE, ISC
129. "Perhaps a variety of the last, with more numerous flowers and larger legumes," **A. sparsiflorus** var. **majusculus,** isotype, !DUKE, ISC
130. **A. bisulcatus,** !DUKE
131. **A. lotiflorus,** isotypes of var. *pedunculosus* and var. *brachypus,* !DUKE, ISC
132. *A. caryocarpus:* **A. crassicarpus,** !DUKE
133. *A. caryocarpus:* **A. crassicarpus,** !DUKE
134. **A. pectinatus,** !DUKE, ISC
135. **Oxytropis splendens,** !DUKE
136. *Astragalus striatus:* **A. adsurgens** var. **robustior,** !DUKE, ISC
137. **A. frigidus** [var. **americanus**], "subalpine, in wet pine woods," !DUKE, ISC [Undoubtedly collected in South Park, this has not been found again in Colorado.]
138. *A. filifolius:* **A. ceramicus,** !DUKE, ISC
139. *A. hypoglottis:* **A. agrestis,** !DUKE, ISC

140. *Oxytropis lambertii:* **O. sericea** and **O. lambertii,** !DUKE, ISC (*O. sericea*) [Gray's notes suggest hybrid swarms involving *O. sericea.* These are common in the area covered.]
141. *Astragalus pauciflorus?:* **A. leptaleus,** !DUKE, ISC
142. *A. decumbens:* **A. miser** subsp. **oblongifolius,** !ISC
143. *Oxytropis arctica:* **O. podocarpa,** !DUKE; **O. parryi,** syntype collection, !DUKE, ISC
144. **O. multiceps,** !DUKE, ISC
145. *Astragalus sericoleucus:* **Orophaca sericea,** !DUKE, ISC [on the plains]
146. *Sophora sericea:* **Vexibia nuttalliana,** !DUKE [on the plains]
147. **Glycyrrhiza lepidota,** !DUKE, ISC
148. *Prunus pensylvanica:* **Cerasus pensylvanica,** !DUKE, ISC
149. *Spiraea dumosa:* **Holodiscus dumosus,** !DUKE, ISC
150. *S. opulifolia* var. *parvifolia:* **Physocarpus monogynus,** !DUKE, ISC
151. **Sibbaldia procumbens,** !DUKE, ISC
152. *Geum triflorum:* **Erythrocoma triflora,** !DUKE, ISC
153. **Dryas octopetala** [subsp. **hookeriana**], !DUKE, ISC
154. *Potentilla fissa:* **Drymocallis fissa,** !DUKE
155. *P. fruticosa:* **Pentaphylloides floribunda,** !DUKE
156. No entry: **Acomastylis rossii** subsp. **turbinata,** !DUKE, ISC
157. **Potentilla concinna,** !DUKE
158. *P. pensylvanica* var. *hippiana:* **P. hippiana,** !DUKE
159. *P. fastigiata:* **a. P. pulcherrima. b. P. diversifolia,** !DUKE
160. *P. nivea:* **P. subjuga,** !DUKE
161. **P. plattensis,** !DUKE
162. *P. pensylvanica* var. *strigosa:* **P. pensylvanica,** !DUKE
163. *Rubus deliciosus:* **Oreobatus deliciosus,** !DUKE, ISC
164. *R. triflorus:* **Cylactis pubescens,** !DUKE, ISC
165. *Cercocarpus parvifolius:* **C. montanus,** !DUKE, ISC
166. **Epilobium palustre** var. **grammadophyllum,** !DUKE, ISC
167. *E. alpinum:* **E. hornemannii,** !DUKE, ISC
168. *E. paniculatum:* **E. brachycarpum,** !DUKE
169. *E. latifolium:* **Chamerion subdentatum,** !DUKE [with one stem of **E. anagallidifolium**]
170. *E. angustifolium:* **Chamerion danielsii,** !DUKE, ISC
171. *Gayophytum racemosum:* **G. diffusum** subsp. **parviflorum,** !DUKE, ISC
172. **G. ramosissimum,** !DUKE
173. *Oenothera marginata:* **O. caespitosa** subsp. **macroglottis,** !DUKE, ISC

174. *O. missouriensis:* **O. howardii,** !ISC
175. *O. triloba:* **O. flava,** !DUKE, ISC
176. *O. nuttallii* (*Taraxia longiflora* and *breviflora*): **O. flava, O. albicaulis,** !DUKE [*O. flava* only], ISC
177. *O. pinnatifida:* **O. albicaulis,** !DUKE, ISC
178. **O. coronopifolia,** !DUKE, ISC
179. *O. serrulata:* **Calylophus serrulatus,** !DUKE, ISC
180. **Gaura parviflora,** !DUKE
181. **G. coccinea,** !DUKE
182. **Hippuris vulgaris,** !DUKE, ISC
183. *Opuntia missouriensis:* **O. polyacantha,** !ISC [one immature pad], DUKE [pad and flower]
184. *Ribes lacustre:* **R. montigenum,** !DUKE [ISC specimen is *R. cereum*]
185. **R. leptanthum,** !DUKE, ISC
186. **R. cereum,** !DUKE
187. *R. hirtellum:* **Ribes inerme,** !DUKE
188. **R. aureum,** !DUKE, ISC
189. *Sedum rhodanthum:* **Clementsia rhodantha,** !DUKE
190. *S. stenopetalum:* **Amerosedum lanceolatum,** !DUKE, ISC
191. *S. rhodiola:* **Rhodiola integrifolia,** !DUKE, ISC
192. **Heliotropium curassavicum** var. **oculatum,** !DUKE, ISC
193. *Saxifraga nivalis:* **Micranthes rhomboidea,** !DUKE [The specimen of this number at ISC is **M. oregana**.]
194. *S. nivalis:* **Micranthes rhomboidea,** !DUKE, ISC [This collection has some plants with branched inflorescences. Such forms are infrequent in the Front Range. A specimen with this number, !ISC, is **Goodyera oblongifolia.**]
195. **S. cernua,** !DUKE, ISC
196. *S. controversa:* **Muscaria adscendens,** !DUKE, ISC
197. *S. bronchialis:* **Ciliaria austromontana,** !DUKE, ISC
198. *S. debilis:* **S. hyperborea** subsp. **debilis,** !DUKE, ISC
199. *S. serpyllifolia:* **Hirculus serpyllifolius** subsp. **chrysanthus,** !DUKE
200. *S. flagellaris:* **Hirculus platysepalus** subsp. **crandallii,** !DUKE
201. *S. hirculus:* **Hirculus prorepens,** !DUKE, ISC
202. **Androsace chamaejasme** [subsp. **carinata**], !DUKE, ISC
203. *Saxifraga jamesii:* **Telesonix jamesii,** !DUKE
204. *S. punctata:* **Micranthes odontoloma** [DUKE specimen is **Heuchera parvifolia.**]
205. *Heuchera bracteata* plus *H. hallii,* isotype of **H. hallii,** !DUKE, ISC
206. *Lithophragma parvifolia:* **L. parviflorum,** !DUKE, ISC
207. No entry: **Micranthes odontoloma,** !DUKE

208. **Mitella pentandra,** !DUKE, ISC
209. No entry: **Tragia ramosa,** !DUKE, ISC
210. *Cymopterus glomeratus:* **C. acaulis,** !DUKE, ISC
211. **C. montanus,** !DUKE
212. *Peucedanum nudicaule?:* **Lomatium orientale,** !DUKE
213. *Cymopterus alpina:* **Oreoxis alpina,** !ISC [DUKE specimen appears to be **O. humilis.**]
214. *Musenium trachyspermum:* **Musineon divaricatum,** !DUKE
215. *Thaspium trachypleurum,* isotype: **Harbouria trachypleura,** !DUKE
216. *Conioselinum fischeri:* **C. scopulorum,** !ISC [DUKE specimen is **Ligusticum tenuifolium.**]
217. *Thaspium montanum:* **Pseudocymopterus montanus,** !DUKE
218. *Conioselinum canadense:* **C. scopulorum,** !ISC [DUKE specimen is **Ligusticum porteri.**]
219. *Archangelica gmelinii:* **Angelica grayi,** !DUKE
220. *Archemora fendleri:* **Oxypolis fendleri,** !DUKE, ISC
221. "acaulescent, unidentifiable for lack of fruit" [Two specimens at DUKE on one sheet. One is **Adoxa moschatellina;** the other is **Aletes acaulis.**]
222. *Cymopterus anisatus,* isotype: **Aletes anisatus,** !DUKE, ISC
223. **Adoxa moschatellina,** !DUKE, ISC
224. **Linnaea borealis** [subsp. **americana**], !DUKE, ISC
225. *Symphoricarpos montanus:* **S. rotundifolius,** !DUKE
226. *Lonicera involucrata:* **Distegia involucrata,** !DUKE
227. **Symphoricarpos occidentalis,** !DUKE, ISC
228. *Viburnum pauciflorum:* **V. edule,** !DUKE, ISC
229. *Galium boreale:* **G. septentrionale,** !DUKE, ISC
230. **G. trifidum** var. **subbiflorum,** !DUKE, ISC
231. *Valeriana dioica* var. *sylvatica:* **V. capitata** subsp. **acutiloba,** !DUKE, ISC
232. *Erigeron acre:* **Trimorpha elongata,** !DUKE, ISC
233. *Diplopappus ericoides:* **Leucelene ericoides,** !DUKE
234. **Erigeron compositus,** !DUKE, ISC [mixture with **E. pinnatisectus**]
235. *E. sp.:* **E. simplex,** !DUKE; **E. vetensis,** !DUKE, ISC
236. **E. glabellus,** !DUKE
237. *E. divergens:* **E. flagellaris,** !DUKE, ISC
238. *E. grandiflorum* var. *elatius:* **E. elatior,** !DUKE
239. *E. glabellus* var. *molle, nomen nudum:* **E. subtrinervis,** !DUKE
240. **E. glabellus,** !DUKE, ISC
241. *Aster salsuginosus:* **Erigeron eximius,** !DUKE
242. *A. glacialis:* **Erigeron leiomerus,** !DUKE

243. *Erigeron uniflorum:* **E. simplex,** !DUKE
244. *E. caespitosus:* **E. canus,** !DUKE, ISC
245. **E. pumilus,** !DUKE, ISC
246. **E. bellidiastrum,** !DUKE [with a stem of **E. divergens;** collected on the plains]
247. *Solidago lanceolata:* **Euthamia occidentalis,** !DUKE, ISC
248. *S. nemoralis:* **S. simplex** var. **nana,** !DUKE, ISC
249. **S. missouriensis,** !DUKE, ISC
250. *S. virga-aurea:* **S. simplex,** !DUKE, ISC
251. *S. virga-aurea* var. *multiradiata:* **S. multiradiata** var. **scopulorum,** !DUKE, ISC
252. *Aster adscendens* var. *ciliatifolius:* **A. occidentalis** var. **fremontii,** !ISC [det. A. G. Jones], DUKE
253. *A. spp.:* **A. foliaceus** var. **apricus,** !ISC [det. A. G. Jones, who saw other sheets of this number and notes that this is a mixed collection], !DUKE [Part of this number was cited in the protologue of *Aster ascendens* var. *parryi.*]
254. *A. ericoides:* **A. porteri** [This is a renaming, at specific rank, of *A. ericoides* var. *strictus*], !DUKE, ISC
255. *Aplopappus inuloides:* **Pyrrocoma clementis,** !DUKE, ISC
256. *A. pygmaeus + Aplopappus lyallii:* **Tonestus pygmaeus,** !DUKE + **Tonestus lyallii,** !GH
257. *A. croceus,* isotype: **Pyrrocoma crocea,** !DUKE, ISC [This a very small specimen.]
258. *A. fremontii:* **Oönopsis foliosa,** !DUKE [collected in the Arkansas River valley]
259. *A. parryi:* **Oreochrysum parryi,** !DUKE
260. *Chrysopsis villosa,* "with the dwarf variety, *C. hispida*": **Heterotheca villosa,** !DUKE, ISC [The "dwarf variety" is probably **H. pumila.**]
261. **Iva axillaris,** !DUKE, ISC
262. *Iva ciliata:* **I. annua,** !DUKE, ISC [Probably collected in Nebraska. This midwestern species does not occur in Colorado.]
263. *Euphrosyne xanthifolia:* **Cyclachaena xanthifolia,** !DUKE, ISC
264. *Franseria tomentosa:* **Ambrosia tomentosa,** !DUKE, ISC
265. *F. hookeriana:* **Ambrosia acanthicarpa,** !ISC
266. *Lepachys columnaris:* **Ratibida columnifera,** !DUKE
267. **Gaillardia aristata,** !DUKE [and a second specimen labeled 276]
268. *Helianthella uniflora:* **H. quinquenervis,** !DUKE, ISC
269. **Helianthus pumilus,** !DUKE, ISC
270. **H. petiolaris,** !DUKE, ISC
271. **Heliomeris multiflora,** !DUKE, ISC

272. *Helenium hoopesii,* isotype: **Dugaldia hoopesii,** !DUKE, ISC
273. *Actinella grandiflora:* **Rydbergia grandiflora,** !DUKE
274. *A. richardsonii:* **Picradenia richardsonii,** !DUKE
275. *A. scaposa:* **Tetraneuris acaulis,** !DUKE
276, 277. *A. acaulis,* "in different forms" [The collections, !DUKE, are **Tetraneuris brevifolia.**]
278. *Bahia oppositifolia:* **Picradeniopsis oppositifolia,** !DUKE, ISC
279. *Thelesperma gracile:* **T. megapotamicum,** !DUKE, ISC
280. **T. filifolium** var. **intermedium,** !DUKE, ISC
281. *Villanova chrysanthemoides:* **Bahia dissecta,** !DUKE
282. *Hymenopappus tenuifolius:* **H. filifolius** subsp. **cinereus,** !DUKE, ISC
283. *Chaenactis achilleifolia:* **C. alpina,** !DUKE, ISC
284. *C. achilleifolia* var. *douglasii:* **C. douglasii,** !ISC
285. **Machaeranthera tanacetifolia,** !DUKE [specimen also with two stems of *M. pattersonii*], ISC [type collection of *Aster pattersonii* var. *hallii*]
286. **Grindelia squarrosa,** !DUKE, ISC [with one stem of **G. subalpina**]
287. *Aplopappus rubiginosus:* **Machaeranthera phyllocephala,** !DUKE, ISC
288. *A. spinulosus:* **Machaeranthera pinnatifida,** !DUKE
289. **Townsendia grandiflora,** !DUKE
290. *T. sericea:* **T. exscapa,** !DUKE, ISC
291. *Aster angustus:* **Brachyactis ciliata** subsp. **angusta,** !ISC, det. A. Jones
292. *Linosyris graveolens:* **Chrysothamnus nauseosus** subsp. **graveolens,** !DUKE [one of two sheets is subsp. **nauseosus**], ISC
293. *L. parryi,* isotype: **Chrysothamnus parryi,** !DUKE, ISC
294. *Gutierrezia euthamiae:* **G. sarothrae,** !DUKE, ISC
295. *Linosyris viscidiflora:* **Chrysothamnus viscidiflorus,** !DUKE
296. **Macronema discoideum,** !DUKE, ISC
297. **Pectis angustifolia,** !DUKE, ISC, "sand hills along the Platte River, on the plains"
298. **Artemisia arctica** [subsp. **saxicola**], !DUKE
299. **A. scopulorum,** isotype, !DUKE, ISC
300. *A. canadensis:* **Oligosporus pacificus,** !DUKE
301. *A. sp.* "a glabrous form of the last": **Oligosporus caudatus,** !DUKE
302. *A. dracunculoides* var. *brevifolia:* **Oligosporus dracunculus** subsp. **glaucus,** !DUKE
303. **A. ludoviciana,** !DUKE
304. **A. frigida,** !DUKE
305. **A. ludoviciana,** !DUKE
306. *A. tridentata:* **Seriphidium vaseyanum,** !DUKE

307. *A. filifolia:* **Oligosporus filifolius,** !DUKE
308. *Obione canescens:* **Atriplex canescens,** !DUKE, ISC
309. *Antennaria carpathica* var. *pulcherrima:* **A. pulcherrima** subsp. **anaphaloides,** !DUKE, ISC [One sheet, !DUKE, is **Anaphalis margaritacea**. In a footnote, Gray proposed the following, *Antennaria margaritacea* var. *subalpina,* based on *Parry 421,* calling it a "singular, nearly alpine form."]
310. *A. dioica,* and *A. alpina,* mixed*:* **A. rosea,** !DUKE [A second sheet of this number contains one plant of *A. rosea* and one of **A. microphylla.**]
311. *Gnaphalium strictum:* **G. uliginosum,** !DUKE, ISC
312. *G. decurrens:* **Pseudognaphalium stramineum,** !DUKE, ISC [with a small stem of **Antennaria media**]
313. **Brickellia grandiflora** var. **minor,** syntype [with *Parry 423*], !DUKE, ISC
314. *Nardosmia sagittata:* **Petasites sagittatus,** !DUKE, ISC
315. **Liatris punctata,** !DUKE, ISC
316. *Senecio lugens:* **S. integerrimus,** !DUKE
317. *S. amplectens:* **Ligularia amplectens,** !DUKE, ISC [The latter sheet contains two stems of *L. amplectens* and one of **L. holmii.**]
318. **S. integerrimus,** !DUKE
319. *S. soldanella,*isotype: **Ligularia soldanella,** !ISC, DUKE
320. *S. cernuus:* **Ligularia pudica,** !DUKE, ISC
321. *S. bigelowii* var. *hallii,* isotype: **Ligularia bigelovii** var. **hallii,** !DUKE
322. **S. fremontii** [subsp. **blitoides**], !DUKE
323. **S. triangularis,** !DUKE, ISC
324. *S. andinus?* [Gray surmised this might be a hybrid of *S. triangularis* and *S. serra.* These do occur, rarely.]
325. *S. lugens,* "the downy state": **S. atratus,** !DUKE
326. *S. sp.* "a dwarf form of" [316]: **S. wootonii,** !DUKE
327. **S. eremophilus** [subsp. **kingii**], !DUKE
328. *S. longilobus* [Gray called this number a mixture of a plains plant and a mountain form. The two specimens at DUKE are **S. spartioides.** That at ISC is **S. flaccidus** subsp. **douglasii.**]
329. *S. canus:* **Packera cana,** !DUKE, ISC
330. No entry: **Packera cana,** !DUKE
331. *S. aureus* var. *(alpinus) werneriaefolius:* **Packera werneriifolia,** isotype of *S. aureus* var. *werneriaefolius,* !DUKE [two sheets, one as 311], ISC
332. *S. aureus?* var. *croceus, nomen nudum:* **Packera crocata,** isotype of *Senecio crocatus* [Gray also mentions a discoid specimen. The specimen at ISC is **P. pauciflora,** which occurs in South Park.]

333. *S. aureus* var. *borealis* and var. *balsamitae,* "Some of the specimens are passing to *S. fendleri.*" [One DUKE specimen is **Packera tridenticulata.** A second sheet contains *P. fendleri, P. crocata,* and *P. tridenticulata.*]
334. *Arnica angustifolia:* **A. fulgens,** !DUKE, ISC [with a stem of **A. rydbergii**]
335. **A. mollis,** !DUKE, ISC
336. **A. cordifolia,** !DUKE
337. **A. chamissonis** [subsp. **foliosa**], !DUKE, ISC
338. *A. angustifolia?* var. *eradiata, nomen nudum,* isotype: **A. parryi,** !DUKE, ISC
339. *Cirsium acaule* var. *americanum,* isotype: **C. scariosum,** !DUKE, ISC
340. *Cnicus edule?:* **C. parryi,** syntype of *Cnicus parryi,* !DUKE, ISC
341. *Cirsium eriocephalum,* isotype: **C. scopulorum,** !DUKE [The specimen seen at ISC appears to be what is being called *C. eatonii.*]
342. *Echinais carlinoides* var. *nutans,* isotype of *Cnicus carlinoides* var. *americanum:* **Cirsium centaureae,** !DUKE, ISC
343. *Cirsium drummondii:* **C. coloradense,** !DUKE, ISC
344. *Mulgedium pulchellum:* **Lactuca tatarica** subsp. **pulchella,** !DUKE
345. **Lygodesmia juncea,** !DUKE
346. *Stephanomeria runcinata:* **S. wrightii,** !DUKE, ISC
347. *Lygodesmia juncea* var? *rostrata,* isotype: **Shinnersoseris rostrata,** !DUKE, ISC
348. *Crepis runcinata:* **Psilochenia runcinata,** !DUKE, ISC
349. *Hieracium triste:* **Chlorocrepis tristis** subsp. **gracilis,** !DUKE
350. *H. albiflorum:* **Chlorocrepis albiflora,** !DUKE, ISC
351. *Nabalus racemosus:* **Prenanthes racemosa,** !DUKE, ISC
352. *Palafoxia hookeriana:* **P. sphacelata,** !DUKE, ISC [on the plains]
353. *Crepis occidentalis:* **Psilochenia occidentalis,** !DUKE
354. *Troximon glaucum:* **Agoseris glauca** var. **dasycephala,** !DUKE
355. *Macrorhynchus troximoides:* **Agoseris aurantiaca,** !DUKE, ISC
356. *Troximon glaucum* var. *dasycephalum:* **Agoseris glauca** var. **dasycephala,** !DUKE, ISC
357. *Taraxacum montanum:* **T. ovinum,** !DUKE, ISC
358. **Campanula rotundifolia,** !DUKE
359. *C. langsdorffiana:* **C. parryi,** !DUKE, ISC [with a few stems of no. 361]
360. **C. uniflora,** !DUKE, ISC
361. **C. aparinoides,** "a depauperate form," !DUKE, ISC [This still remains the only known collection from Colorado. It was undoubtedly collected in meadows of South Park and was not uncommon, judging from the full sheets.]

362. **Vaccinium myrtillus** [subsp. **oreophilum**], !DUKE [with **V. scoparium**], ISC
363. **V. cespitosum,** !DUKE, ISC
364. **Arctostaphylos uva-ursi,** !DUKE [with small amount of *Vaccinium cespitosum*]
365. *Gaultheria myrsinites:* **G. humifusa,** !DUKE
366. *Pyrola secunda:* **Orthilia secunda** subsp. **obtusata,** !DUKE, ISC
367. *P. rotundifolia* var. *uliginosa:* **P. rotundifolia** subsp. **asarifolia,** !DUKE, ISC
368. *P. chlorantha:* **P. minor,** !DUKE, ISC
369. *P. uniflora:* **Moneses uniflora,** !DUKE
370. *Kalmia glauca:* **K. microphylla,** !ISC
371. **Pterospora andromedea**
372. **Plantago eriopoda,** !DUKE, ISC
373. *P. eriopoda var.:* **P. tweedyi,** !DUKE, ISC
374. **P. patagonica,** !DUKE, ISC
375. **Androsace filiformis,** !DUKE, ISC [Gray correctly noted that this is also a Siberian species.]
376. **A. septentrionalis,** !DUKE, ISC
377. **A. occidentalis,** !DUKE, ISC [collected on the plains]
378. *Primula farinosa:* **P. incana,** !DUKE, ISC
379. **P. parryi,** !DUKE
380. **P. angustifolia,** !DUKE, ISC
381. *Dodecatheon meadia:* **D. pulchellum,** !DUKE, ISC
382. **Lysimachia ciliata,** !DUKE, ISC
383. **Aphyllon fasciculatum,** !DUKE
384. *Penstemon glaber:* **P. brandegei,** !DUKE, ISC [probably collected near Colorado Springs]
385. *P. acuminatus:* **P. virgatus** subsp. **asa-grayi,** !DUKE, ISC
386. *P. acuminatus:* **P. secundiflorus,** !DUKE, ISC
387. *P. humilis:* **P. virens,** !DUKE, ISC
388. **P. hallii,** isotype, !DUKE, ISC
389. **P. albidus,** !DUKE, ISC [collected on the plains]
390. *P. acuminatus:* **P. angustifolius,** !DUKE, ISC
391. *P. confertus* var. *purpureo-coeruleus:* **P. procerus,** !DUKE, ISC
392. *P. glaucus?* var. *stenosepalus:* **P. whippleanus,** !DUKE, ISC
393. **P. caespitosus,** !DUKE, ISC [the erect form, evidently collected in South Park, near Como]
394. *P. pubescens* var. *gracilis:* **P. gracilis,** !DUKE, ISC
395. **P. barbatus** [subsp. **torreyi**], !DUKE
396. **P. harbourii,** isotype, !DUKE, ISC

397. **Chionophila jamesii,** !DUKE
398. *Mimulus luteus* var. *alpinus:* isotype of *M. hallii* (= **M. guttatus** var. **hallii**), !DUKE [with *M. glabratus* var. *fremontii*], ISC
399. *M. jamesii* var. *fremontii:* **M. glabratus** var. **fremontii,** !DUKE, ISC
400. **M. floribundus,** !DUKE, ISC
401. *M. rubellus:* **M. breweri,** !DUKE, ISC
402. **Collinsia parviflora,** !DUKE, ISC
403. *Collomia gracilis:* **Microsteris gracilis,** !DUKE, ISC
404. **C. linearis,** !DUKE
405. *Synthyris plantaginea,* "with a little *S. alpina*": **Besseya plantaginea,** !DUKE, **B. alpina**
406. *Veronica serpyllifolia:* **Veronicastrum serpyllifolium** subsp. **humifusum,** !DUKE
407. *V. alpina:* **V. nutans,** !DUKE
408. **V. americana,** !DUKE
409. *Castilleja breviflora:* **C. puberula,** !DUKE
410. **C. integra,** !DUKE, ISC
411. *C. pallida* var. *miniata:* **C. miniata,** !DUKE [with one stem of *C. rhexifolia;* Gray's comments imply that the collection also includes what we now recognize as *C. rhexifolia.* In fact, the specimen at ISC is that!]
412. *C. pallida:* **C. sulphurea** and **C. occidentalis,** !DUKE, ISC
413. **Orthocarpus luteus,** !DUKE
414. **Pedicularis racemosa** [subsp. **alba**], !DUKE
415. **P. crenulata,** !DUKE, ISC
416. **P. canadensis** [subsp. **fluviatilis**], !DUKE, ISC
417. **P. bracteosa** [subsp. **paysoniana**], !DUKE
418. **P. procera,** !DUKE
419. **P. groenlandica,** !DUKE
420. **P. parryi,** !DUKE
421. *P. sudetica:* **P. scopulorum,** !DUKE
422. *Rhinanthus crista-galli* var. *minor:* **R. minor** subsp. **borealis,** !DUKE, ISC
423. **Hedeoma hispidum,** !DUKE, ISC [on the plains]
424. **H. drummondii,** !DUKE, ISC [on the plains]
425. *Mentha canadensis* var. *glabrata:* **M. arvensis,** !DUKE, ISC
426. *Salvia trichostemoides:* **S. reflexa,** !DUKE
427. *S. pitcheri:* **S. azurea** var. **grandiflora,** !DUKE [collected on the plains]
428. *Monarda aristata:* **M. pectinata,** !DUKE, ISC

429. *Lophanthus anisatus:* **Agastache foeniculum,** !DUKE, ISC [undoubtedly collected in the lower Platte Canyon; still known from a very few Colorado localities]
430. **Dracocephalum parviflorum,** !DUKE, ISC
431. *Scutellaria resinosa:* **S. brittonii,** !DUKE
432. **S. galericulata** var. **epilobiifolia,** !DUKE
433. *Echinospermum redowskii,* "and a depauperate, diffuse, or procumbent form of *Eritrich*[*i*]*um californicum*": **Lappula redowskii,** !DUKE, and **Plagiobothrys scouleri** subsp. **penicillata,** !DUKE, ISC
434. *Eritrich*[*i*]*um crassisepalum* and "a more upright and narrower-leaved species," a mixture of **Cryptantha crassisepala, C. fendleri,** and **Oreocarya thyrsiflora,** !DUKE, ISC [*C. crassisepala* only]
435. *E. jamesii:* **Oreocarya suffruticosa,** !DUKE, ISC
436. *Heliotropium convolvulaceum:* **Euploca convolvulacea,** !DUKE [collected on the plains]
437. *Echinospermum floribundum:* **Hackelia floribunda,** !DUKE, ISC
438. *Eritrich*[*i*]*um glomeratum,* "and a form with shorter and more branched inflorescence": **Oreocarya virgata,** !DUKE, and **O. thyrsiflora**
439. *Phacelia circinata:* **P. heterophylla,** !DUKE
440. **Eritrich[i]um aretioides,** !DUKE, ISC
441. *Lithospermum pilosum:* **L. multiflorum,** !DUKE
442. *Mertensia sibirica:* **M. ciliata,** !DUKE
443. *M. alpina:* **M. lanceolata** var. **viridis,** !DUKE
444. **M. alpina,** !DUKE, and **M. lanceolata,** the **"bakeri"** race, !ISC
445. *M. alpina* var.: **M. lanceolata,** !DUKE, ISC
446. *Phacelia popei:* **P. alba,** !DUKE
447. **P. sericea,** !DUKE, ISC
448. *Polemonium coeruleum,* "a very viscid and glandular variety": **P. foliosissimum,** !DUKE
449. **P. coeruleum** [subsp. **amygdalinum**], !DUKE [The ISC specimen belongs to *P. foliosissimum;* mounted with this is *Parry 275.*]
450. **P. confertum,** isotype, !DUKE [with **P. viscosum**], ISC
451. *P. confertum* var. *mellitum,* isotype: **P. brandegei,** !DUKE, ISC [as no. 417. See Grant, V. Bot. Gaz. 150:158–169.1989, for discussion of these specimens.]
452. *P. pulchellum:* **P. pulcherrimum** subsp. **delicatum,** !DUKE
453. *Phlox douglasii:* **P. multiflora,** !DUKE, ISC
454. *P. humilis?:* **P. sibirica** subsp. **pulvinata,** !DUKE
455. *P. hoodii:* **P. condensata,** !DUKE
456. **Gilia pinnatifida,** !DUKE

457. *G. inconspicua:* **G. ophthalmoides,** !DUKE, ISC
458. *G. longiflora:* **Ipomopsis laxiflora,** !DUKE, ISC [collected on the plains]
459. *G. aggregata:* **Ipomopsis aggregata** subsp. **collina,** x **I. aggregata** subsp. **candida,** !DUKE [The ISC specimen contains two taxa: *I. aggregata* subsp. *weberi* and *I. aggregata* subsp. *collina* or *candida.*]
460. *G. spicata:* **Ipomopsis spicata,** !DUKE
461. *G. congesta* var? "with the leaves mostly entire. Alpine": **Ipomopsis globularis,** isotype of *Gilia globularis,* !DUKE [undoubtedly from Georgia Pass]
462. **Chamaerhodos erecta** [subsp. **nuttallii**], !DUKE, ISC
463. *Gilia pungens:* **Leptodactylon pungens,** !DUKE
464. *Cuscuta arvensis* var. *pentagona:* **Grammica cuspidata,** !ISC [on *Cycloloma atriplicifolia*], !DUKE [on *Psoralidium*]
465. **Solanum rostratum,** !DUKE
466. *Physalis lobata:* **Quincula lobata,** !DUKE, ISC
467. **Solanum triflorum,** !DUKE, ISC
468. *Gentiana affinis:* **Pneumonanthe afffinis,** !DUKE, ISC
469. *G. affinis* var. *brachycalyx,* isotype: **Pneumonanthe affinis,** !DUKE, ISC
470. *G. parryi:* **Pneumonanthe parryi,** !DUKE, ISC
471. *G. detonsa:* **Gentianopsis thermalis,** !DUKE
472. *G. frigida* var. *algida:* **Gentianodes algida,** !DUKE
473. *G. acuta:* **Gentianella acuta,** !DUKE, ISC [with **G. strictiflora**]
474. *G. humilis:* **Chondrophylla aquatica,** !DUKE, ISC
475. *G. prostrata* var. *americana:* **Chondrophylla prostrata** + **C. nutans,** !DUKE, ISC
476. **Swertia perennis,** !DUKE, ISC [together with one small stem of **Asclepias uncialis**]
477. *Pleurogyne rotata:* **Lomatogonium rotatum** subsp. **tenuifolium,** !DUKE, ISC
478. *Asclepias brachystephana:* **A. uncialis,** !DUKE, ISC
479. **A. speciosa,** !DUKE, ISC
480. *A. ovalifolia* var.: **A. hallii,** isotype, !DUKE, ISC
481. *A. verticillata:* **A. pumila,** !DUKE, ISC
482. *Oxybaphus angustifolius:* **O. linearis,** !DUKE, ISC
483. *O. nyctagineus:* **O. hirsutus,** !DUKE, ISC
484. *Obione argentea:* **Atriplex argentea,** !DUKE, ISC
485. *Chenopodium hybridum:* **C. simplex,** !DUKE
486. **Monolepis nuttalliana,** !DUKE, ISC
487. *Froelichia floridana:* **F. gracilis,** !DUKE

488. *Chenopodina depressa:* **Suaeda calceoliformis,** !DUKE [with **S. nigra**]
489. *C. maritima* var. *erecta:* **Suaeda nigra,** !DUKE
490. *Polygonum bistorta* var. *oblongifolium:* **Bistorta bistortoides,** !DUKE, ISC
491. *P. viviparum:* **Bistorta vivipara,** !DUKE [with **Polygonum polygaloides** subsp. **kelloggii**]
492. *P. tenue:* **P. douglasii,** !DUKE [with **P. ramosissimum** and **P. engelmannii.** The specimen at ISC is *P. engelmannii.*]
493. *P. coarctatum* var. *minus:* **P. polygaloides** subsp. **kelloggii,** !DUKE, ISC
494. **Oxyria digyna,** !DUKE
495. **Rumex venosus,** !DUKE, ISC
496. *R. salicifolius:* **R. triangulivalvis,** !DUKE, ISC
497. *R. maritimus?:* **R. maritimus** subsp. **fueginus,** !DUKE, ISC
498. *R. salicifolius:* **R. triangulivalvis,** !DUKE, ISC
499. *R. longifolius:* **R. aquaticus** subsp. **occidentalis,** !DUKE, ISC
500. *Eriogonum alatum:* **Pterogonum alatum,** !DUKE
501. **E. annuum,** !DUKE
502. **E. effusum,** !DUKE
503. **E. cernuum,** !DUKE, ISC
504. *E. umbellatum,* "straw-colored flowers": **E. subalpinum,** !DUKE; "bright yellow flowers": **E. umbellatum,** !DUKE
505. *E. flavum,* "alpine": **E. flavum,** var. **chloranthum,** !DUKE; "lower elevations," **E. flavum,** !DUKE
506. **Shepherdia canadensis,** !DUKE, ISC
507. *Comandra pallida* var. *angustifolia:* **C. umbellata** subsp. **pallida,** !DUKE, ISC
508. *Euphorbia marginata:* **Agaloma marginata,** !DUKE
509. *E. montana:* **Tithymalus brachyceras,** !DUKE, ISC
510. *E. dictyosperma:* **Tithymalus spathulatus,** !DUKE, ISC
511. *E. hexagona:* **Zygophyllidium hexagonum,** !DUKE
512. *E. petaloidea:* **Chamaesyce missurica,** !DUKE
513. *E. fendleri:* **Chamaesyce fendleri,** !DUKE, ISC
514. *Croton muricatum:* **Croton texensis,** !DUKE
515. *Quercus douglasii* var. *neomexicana:* **Q. gambelii,** !DUKE, ISC [The handwritten "Pine Valley" (Utah?) label may be irrelevant, since both leaf forms on the sheet occur in populations around Pueblo, CO.]
516. *Corylus rostrata:* **C. cornuta,** !DUKE, ISC
517. **Betula glandulosa,** !DUKE
518. *B. papyracea* var.: **B. fontinalis,** !DUKE, ISC

519. *Alnus viridis:* **A. incana** subsp. **tenuifolia,** !DUKE, ISC
520. **Salix arctica** [subsp. **petraea**], !DUKE, ISC
521. **S. reticulata** [subsp. **nivalis**], !DUKE, ISC
522. *S. rostrata:* **S. bebbiana,** !DUKE, ISC
523. *S. glauca:* **S. brachycarpa,** !DUKE, ISC
524. *S. cordata:* **S. monticola,** !DUKE [with a twig of **Populus balsamifera**], !ISC, **S. ligulifolia,** annotated by C. R. Ball as **S. lutea**
525. **Populus angustifolia,** !DUKE, ISC
526. *P. balsamifera* var. *candicans:* **P. balsamifera,** !DUKE, ISC
527. **P. tremuloides,** !DUKE
528. **Pinus ponderosa** [subsp. **scopulorum**], !DUKE
529. **P. flexilis,** !DUKE
530. **P. aristata,** isotype, !DUKE [See Engelmann 1866.]
531. **P. contorta** var. **latifolia,** !DUKE, ISC [as no. 574]
532. **P. edulis,** !DUKE
533. *Abies menziesii:* **Picea pungens;** a second sheet is **P. engelmannii,** !DUKE, ISC [*P. engelmannii* only]
534. *A. douglasii:* **Pseudotsuga menziesii,** !DUKE, ISC
535. *Platanthera hyperborea:* **Limnorchis hyperborea,** !DUKE; **L. stricta,** !ISC
536. *P. obtusata:* **Lysiella obtusata,** !DUKE
537. *Calypso borealis:* **C. bulbosa,** !DUKE
538. *Cypripedium parviflorum:* **C. calceolus** subsp. **parviflorum,** !DUKE, ISC
539. *Spiranthes gemmipara,* "mixed (in my set) with taller specimens, from the plains, of *S. cernua*": **S. romanzoffiana,** !DUKE, ISC; **S. magnicamporum,** collected in Nebraska, according to Sheviak [personal communication]
540. **Triglochin palustre,** !DUKE, ISC
541. **T. maritimum,** !DUKE, ISC
542. *Iris tenax?:* **I. Missouriensis,** !DUKE, ISC
543. *Streptopus amplexifolius:* **S. fassettii,** !DUKE
544. *Smilacina stellata:* **Maianthemum stellatum,** !DUKE
545. *Allium stellatum:* **A. textile,** !DUKE, ISC
546. *A. schoenoprasum:* **A. geyeri,** !DUKE, ISC
547. **A. cernuum,** !DUKE, ISC
548. **Leucocrinum montanum,** !DUKE, ISC
549. *Calochortus venustus:* **C. gunnisonii,** !DUKE
550. *Zygadenus glaucus:* **Anticlea elegans,** !DUKE
551. *Amianthium nuttallii:* **Toxicoscordion venenosum,** !DUKE, ISC
552. **Lloydia serotina,** !DUKE, ISC

553. **Frasera speciosa,** !DUKE
554. **Luzula spicata,** !DUKE, ISC
555. **L. parviflora,** !DUKE, ISC
556. **L. comosa,** !DUKE, ISC
557. *Juncus triglumis:* **J. albescens,** !DUKE, ISC
558. *J. articulatus* var. *pelocarpus:* **J. alpino-articulatus,** !DUKE, ISC
559. **J. bufonius,** !DUKE, ISC
560. **J. castaneus,** !DUKE, ISC
561. *J. arcticus* var. *gracilis?:* **J. parryi,** syntype, !DUKE, ISC
562. *J. arcticus:* **J. hallii,** isotype, !DUKE, ISC
563. *J. arcticus:* **J. drummondii,** !DUKE, **J. parryi,** !ISC
564. *J. xiphioides:* **J. mertensianus,** !ISC [The specimen at DUKE is **J. saximontanus.**]
565. *J. ensifolius:* **J. mertensianus,** !DUKE
566. *J. menziesii:* **J. longistylis,** !DUKE, ISC
567. *J. balticus:* **J. arcticus** subsp. **ater,** !DUKE, ISC
568. **Jamesia americana,** !DUKE, ISC
569. *Mentzelia nuda:* **Nuttallia nuda,** !DUKE, ISC
570. *M. multiflora:* **Nuttallia cf. Speciosa,** !ISC
571. *M. albicaulis,* "and some *M. oligosperma,*" !DUKE [The ISC specimen is **Acrolasia albicaulis;** a specimen at DUKE is **A. dispersa** (probably the plant referred to by Gray as **Mentzelia oligosperma**).]
572. **Abronia fragrans,** !DUKE
573. *A. cycloptera:* **Tripterocalyx micranthus,** !DUKE
574. *Arceuthobium campylopodum:* **A. americanum,** on *Pinus contorta,* !DUKE, ISC
575. **Parnassia parviflora,** !DUKE, ISC
576. *Chrysosplenium alternifolium:* **C. tetrandrum,** !DUKE, ISC
577. **Glaux maritima** var. **angustifolia,** !DUKE, ISC
578. **Parnassia fimbriata,** !DUKE, ISC
579. *Evolvulus argenteus:* **E. nuttallianus,** !DUKE, ISC [on the plains]
580. *Utricularia vulgaris?:* **U. ochroleuca,** !DUKE [This occurs in South Park.]
581. *Fimbristylis laxa:* **F. puberula** var. **interior,** !DUKE, ISC [collected on the plains, not in Colorado]
582. *Scirpus pauciflorus* Lightf.: **Eleocharis quinqueflora,** !DUKE, ISC
583. *S. caespitosus:* All specimens, including that reported by C. B. Clarke, !G, belong to **Trichophorum pumilum,** !DUKE, ISC
584. *Cyperus schweinitzii:* **Mariscus schweinitzii,** !ISC [An unnumbered specimen at DUKE is **M. fendlerianus.**]

585. *Carex atrata:* **C. bella,** !DUKE, ISC
586. *C. atrata:* **C. chalciolepis,** !DUKE, ISC
587. *C. atrata:* **C. nova** L. H. Bailey, !DUKE, ISC [type collection of *C. violacea,* erroneously attributed to California]
588. *C. atrata* and *C. rigida:* **C. epapillosa,** and **C. nebrascensis,** !DUKE; **C. illota,** and **C. nebrascensis,** !ISC
589. *C. festiva:* **C. ebenea,** !DUKE, ISC [Specimen at ISC also contains **C. festivella.**]
590. *C. festiva:* **C. microptera,** !DUKE, ISC [+ **C. festivella**]
591. *C. bonplandii* var. *minor,* isotype, = **C. illota,** !DUKE, ISC
592. *C. muricata:* **C. cf. haydeniana**, !DUKE, ISC [A specimen at DUKE is **C. occidentalis.**]
593. *C. siccata:* **C. foenea,** !ISC
594. *C. disticha:* **C. praegracilis,** !DUKE, ISC
595. *C. gayana:* **C. simulata,** !DUKE, ISC [+ **C. stenophylla** subsp. **eleocharis**]
596. **C. deweyana,** !DUKE, ISC
597. **C. stenophylla** [subsp. **eleocharis**], !DUKE
598. *Kobresia scirpina:* **Kobresia myosuroides,** !DUKE, ISC
599. *K. scirpina:* **Kobresia simpliciuscula,** !ISC
600. *Carex douglasii,* !DUKE, ISC
601. *C. tenella:* **C. disperma,** !DUKE, ISC
602. **C. canescens,** !ISC
603. *C. polytrichoides:* **C. leptalea,** !DUKE, ISC
604. *C. filifolia* var.: **C. oreocharis,** isotype, !ISC
605. **C. filifolia,** !DUKE, ISC
606. **C. obtusata,** !ISC
607. *C. pauciflora:* **C. microglochin,** !ISC
608. *C. pyrenaica:* **C. crandallii,** !DUKE, ISC
609. **C. nigricans,** !DUKE, ISC
610. **C. scirpoidea,** !DUKE, ISC [with **C. parryana**]
611. **C. geyeri,** !ISC
612. *C. backii:* **C. saximontana,** !DUKE, ISC
613. **C. capillaris,** !DUKE, ISC
614. *C. longirostris* var. *minor:* **C. sprengelii,** !DUKE, ISC
615. *C. ampullacea:* **C. utriculata,** !ISC
616. *C. jamesii* and *C. angustata:* **C. aquatilis, C. nebrascensis, Juncus drummondii,** !DUKE

616a. *C. angustata:* **C. aquatilis,** !ISC [This sheet bears no number; it has a Parry label with his name crossed out and replaced by *Hall & Harbour.*]

617. **C. parryana** [subsp. **hallii**], isotype, *C. hallii,* !DUKE, ISC [*C. parryana* was named for Sir William Edward Parry!]
618. *C. alpina:* **C. norvegica** subsp. **stevenii,** !ISC
619. **C. buxbaumii,** !DUKE, ISC
620. **C. rossii,** !DUKE, ISC
621.?: **Festuca hallii,** isotype of *Melica hallii,* !DUKE, ISC
622. *Danthonia sericea:* **D. parryi,** isotype, !DUKE, ISC [with *D. intermedia*]
623. *Avena striata:* **Schizachne purpurascens,** !DUKE, ISC
624. *Calamagrostis sylvatica:* **C. stricta,** !DUKE, **C. purpurascens,** !ISC
625. *Trisetum subspicatum:* **T. spicatum** subsp. **congdonii,** !DUKE, ISC
626. *Stipa viridula:* **Achnatherum lettermanii,** !DUKE, ISC
627. *Aira caespitosa:* **Deschampsia cespitosa,** !DUKE, ISC
628. *Hierochloë borealis:* **H. hirta** subsp. **arctica,** !DUKE, ISC
629. *Glyceria aquatica:* **G. grandis,** !DUKE, ISC
630. *G. airoides:* **Puccinellia airoides,** !DUKE, ISC
631. *Vilfa tricholepis:* **Blepharoneuron tricholepis,** !DUKE, ISC
632. **Muhlenbergia pungens,** isotype, !DUKE
633. *Eriocoma cuspidata:* **Achnatherum hymenoides,** !DUKE
634. **Oryzopsis micrantha,** !DUKE, ISC
635. *Graphephorum? Flexuosum,* isotype: **Redfieldia flexuosa,** !DUKE
636. *Bouteloua oligostachya:* **Chondrosum gracile,** !DUKE, ISC
637. **Buchloë dactyloides,** !DUKE
638. *Munroa squarrosa:* **Monroa squarrosa,** !DUKE
639. **Spartina gracilis,** !DUKE, ISC
640. *Brizopyrum spicatum* var. *strictum:* **Distichlis stricta**
641. *Sporobolus asperifolius:* **Muhlenbergia asperifolia,** !DUKE, ISC
642. *Muhlenbergia gracillima:* **M. torreyi** [The DUKE specimen is **Distichlis stricta** (probably a switched label).]
643. *Sporobolus ramulosus:* **Muhlenbergia minutissima,** !DUKE, ISC [on the plains]
644. *Leptochloa fascicularis:* **Diplachne fascicularis,** !DUKE
645. *Tricuspis purpurea:* **Triplasis purpurea,** !ISC [on the plains]
646. *Stipa mongolica:* **Ptilagrostis porteri,** isotype of *Stipa porteri,* !DUKE, ISC
647. **Sporobolus airoides,** !DUKE, ISC
648. **S. cryptandrus,** !ISC
649. **Calamagrostis stricta,** !DUKE, ISC
650. *Koeleria cristata:* **K. macrantha,** !DUKE, ISC
651. *Andropogon argenteus:* **Bothriochloa laguroides** subsp. **torreyana,** !DUKE [on the plains; at ISC, this number is **Andropogon hallii,** isotype, collected in Nebraska]

652. **Aristida purpurea,** !DUKE, ISC [on the plains]
653. **Paspalum setaceum,** !ISC [probably collected on the Nebraska plains]
654. *Elymus condensatus* and *E. triticoides* intermixed: **Leymus cinereus,** !DUKE, ISC [with *Pascopyrum smithii*]
655. *Triticum repens* var.: **Pascopyrum smithii,** !DUKE, ISC [specimens with different awn lengths]
656. *T. caninum* and *T. repens:* **Elymus trachycaulus,** !ISC
657. *T. aegilopoides:* **Elymus lanceolatus,** !DUKE, ISC
658. *Beckmannia erucaeformis:* **B. syzigachne** subsp. **baicalensis,** !DUKE, ISC
659. **Sporobolus airoides,** !DUKE, ISC
660. *Vilfa depauperata:* **Muhlenbergia richardsonii,** !DUKE, ISC
661. *V. cuspidata:* **Muhlenbergia cuspidata,** !DUKE, ISC
662. *Glyceria pauciflora:* **Torreyochloa pauciflora,** !DUKE, ISC
663. **Catabrosa aquatica,** !DUKE, ISC
664. *Muhlenbergia gracilis:* **M. montana,** !DUKE, ISC [A second sheet at ISC quite certainly represents **M. sylvatica,** which has never been found in Colorado. The specimen could have come from eastern Nebraska, where it is within range.]
665. *Festuca ovina* var. *duriuscula:* **F. saximontana,** !DUKE, ISC
666. *F. rubra:* **F. brachyphylla** subsp. **coloradensis,** !DUKE, ISC
667. *F. scabrella:* **F. thurberi,** !DUKE, ISC
668. *Poa cf. nemoralis:* **Trisetum wolfii,** !DUKE
669. *P. andina:* **P. arida,** !DUKE [The specimen is merely the upper part of a culm.]
670. **P. arctica** and *P. alpina,* !DUKE, ISC [no *P. alpina* on this sheet]
671. *Agrostis varians:* **A. variabilis,** !DUKE [The specimens mentioned with long awns are **A. mertensii,** !DUKE], **A. thurberiana**, !ISC
672. *Poa serotina:* **P. nemoralis** subsp. **interior,** !DUKE, ISC
673. *Agrostis cf. rupestris:* **A. variabilis**
674. *Poa alpina:* **P. fendleriana,** !DUKE, ISC
675. *P. sp.:* **P. cf. arida,** !DUKE. **P. glaucifolia,** !ISC
676. *P. arctica?:* **P. reflexa,** !DUKE, **P. arctica,** !ISC
677. *P. andina:* **P. glauca** subsp. **rupicola,** !ISC
678. *P. sp.,* !DUKE [unidentifiable without basal parts], **P. glaucifolia,** !ISC
679. *Sitanion elymoides:* **Elymus longifolius,** !DUKE, ISC
680. *Triticum caninum* var.: **Elymus scribneri,** !DUKE, ISC
681. *Hordeum jubatum:* **Critesion jubatum,** !DUKE, ISC
682. *Alopecurus pratensis* var. *alpestris:* **A. alpinus** subsp. **glaucus,** !DUKE, ISC

683. *A. geniculatus* var. *aristulatus:* **A. aequalis,** !DUKE, ISC
684. No entry: **Phleum commutatum,** !DUKE
685. *Vaseya comata,* isotype: **Muhlenbergia andina,** !DUKE, ISC
686. *Lepturus paniculatus:* **Schedonnardus paniculatus,** !DUKE [on the plains]
687. *Aspidium filix-mas:* **Dryopteris filix-mas,** !DUKE
688. **Cryptogramma acrostichoides,** !DUKE
689. **Asplenium septentrionale,** !DUKE, ISC
690. *Cystopteris fragilis* + *Woodsia sp.:* **C. fragilis** + **W. scopulina,** !DUKE, ISC [*Cystopteris only*]
691. **Cheilanthes fendleri,** !DUKE
692. **Asplenium trichomanes,** !DUKE, ISC
693. *Nothochlaena fendleri:* **Argyrochosma fendleri**
694. *Polypodium vulgare:* **Polypodium saximontanum,** !ISC
695. *P. dryopteris:* **Gymnocarpium dryopteris**

Echinocactus simpsonii var. *minor,* isotype: **Pediocactus simpsonii** var. **minor**

Gymnosteris parvula, !ISC [On the route taken, this could have been collected near Dillon, in the Blue River Valley of Summit County.]

Parry's Collections From 1862

Parry collected together with Hall & Harbour on their 1862 expedition, as evidenced by the itinerary. However, he collected relatively few specimens for himself, compared to the thousands accumulated by Hall & Harbour and later distributed to Gray, with sets being sold to a number of collectors and institutions. The following numerical list may be incomplete because Parry might have added his collections to the sets being distributed by Hall & Harbour.

002. **Arnica cordifolia,** !ISC [cf. *Hall & Harbour 336*]
004. *Deyeuxia neglecta:* **Calamagrostis stricta,** !ISC [cf. *Hall & Harbour 624*]
027. **Viola nuttallii,** !ISC [mounted with *Hall & Harbour 55*]
038. **Arnica latifolia,** !ISC, "Berthoud Pass"
078. *Ranunculus cymbalaria:* **Halerpestes cymbalaria** subsp. **saximontana,** !ISC [cf. *Hall & Harbour 11*]
083. *Trollius laxus:* **T. albiflorus,** !ISC [cf. *Hall & Harbour 22*]
088. *Anemone patens var.:* **Pulsatilla patens** subsp. **multifida,** !ISC [cf *Hall & Harbour 4*]

120. *Bahia chrysanthemoides:* **B. dissecta,** !ISC [cf. *Hall & Harbour 281*]
144. **Ceanothus fendleri,** !ISC [cf. *Hall & Harbour 90*]
158. *Gentiana detonsa:* **Gentianopsis thermalis,** !ISC [cf. *Hall & Harbour 471*]
177. **Trifolium nanum,** !ISC [cf. *Hall & Harbour 99*]
306. *Gentiana prostrata* var. *americana:* **Chondrophylla prostrata** and **C. nutans,** !ISC [cf. *Hall & Harbour 475*]
346. **Lloydia serotina,** !ISC [cf. *Hall & Harbour 552*]
421. *Antennaria margaritacea* var. *subalpina,* isotype
427. **Parnassia parviflora,** !ISC [cf. *Hall & Harbour 575*]
428. **P. fimbriata,** !ISC [cf. *Hall & Harbour 578*]
430. *Sidalcea malvaeflora:* **S. neomexicana,** !ISC
434. *Trifolium longipes:* **T. rusbyi,** !ISC
435. *Astragalus decumbens:* **A. miser** subsp. **oblongifolius,** !ISC [cf. *Hall & Harbour 142*]
436. **Gaura parviflora: Gaura mollis,** !ISC [cf. *Hall & Harbour 180*]
437. *Cornus canadensis:* **Chamaepericlymenum canadense**
441. *Spiranthes gemmipara:* **S. romanzoffiana,** !ISC [cf. *Hall & Harbour 539*]

Aconitum columbianum, !ISC [cf. *Hall & Harbour 29*]

Agrimonia eupatoria: **A. striata,** !ISC

Arceuthobium americanum, !ISC [cf *Hall & Harbour 574*]

Aster (Machaeranthera) coloradoensis: **Machaeranthera pattersonii,** !ISC [cf. *Hall & Harbour 285*]

Astragalus hallii, !ISC [cf. *Hall & Harbour 121*]

Astragalus tridactylicus: **A. missouriensis,** !ISC [cf. *Hall & Harbour 127*]

Astragalus shortianus, !ISC [cf. *Hall & Harbour 126*]

Bigelowia howardii, isotype: **Chrysothamnus parryi** subsp. **howardii,** !ISC, "Middle Park"

Buchloë dactyloides, !ISC [cf. *Hall & Harbour 637*]

Calamagrostis sylvatica: **C. purpurascens,** !ISC [cf. *Hall & Harbour 624*]

Caltha leptosepala: **Psychrophila leptosepala,** !ISC [cf. *Hall & Harbour 21*]

Calypso borealis: **C. bulbosa,** !ISC [cf. *Hall & Harbour 537*]

Campanula aparinoides, !ISC [cf. *Hall & Harbour 361*]

C. scheuchzeri: **C. parryi,** !ISC [cf. *Hall & Harbour 359*]

Clematis ligusticifolia, !ISC [cf. *Hall & Harbour 3*]

Corydalis aurea var. *occidentalis:* **C. aurea** var. **aurea.,** !ISC [cf. *Hall & Harbour 31*]

Circaea alpina [subsp. **pacifica**], !ISC

Epilobium alpinum: **E. anagallidifolium,** !ISC [cf. *Hall & Harbour 169*]
E. angustifolium: **Chamerion danielsii,** !ISC [cf. *Hall & Harbour 169* in part]
E. latifolium: **Chamerion subdentatum,** !ISC [cf. *Hall & Harbour 170*]
Erigeron armeriaefolius: **Trimorpha lonchophylla,** !ISC
Gentiana acuta: mix, **Gentianella acuta** and **G. strictiflora,** !ISC [cf. *Hall & Harbour 473*]
G. amarella: **Gentianella acuta,** !ISC
G. humilis: **Chondrophylla aquatica,** !ISC [cf. *Hall & Harbour 474*]
Geranium: **G. caespitosum,** !ISC [cf. *Hall & Harbour 89*]
Geum rossii: **Acomastylis rossii** subsp. **turbinata,** !ISC [cf. *Hall & Harbour 156*]
G. macrophyllum var. **perincisum,** !ISC
Glaux maritima, !ISC [cf. *Hall & Harbour 60*]
Helianthella: **H. quinquenervis,** !ISC, "Middle Park" [cf. *Hall & Harbour 268*]
H. parryi, isotype, !ISC, "base of Pike's Peak"
Helenium hoopesii: **Dugaldia hoopesii,** !ISC [cf. *Hall & Harbour 272*]
Hieracium albiflorum: **Chlorocrepis albiflora,** !ISC [cf. *Hall & Harbour 350*]
Jamesia americana, !ISC [cf. *Hall & Harbour 568*]
Leucocrinum montanum, !ISC [cf. *Hall & Harbour 548*]
Lilium philadelphicum, !ISC
Lindernia dubia var. **anagallidea,** !ISC
Mimulus rubellus: **M. breweri,** !ISC [cf. *Hall & Harbour 401*]
Oenothera coronopifolia, !ISC [cf. *Hall & Harbour 178*]
Oxytropis nana: **O. multiceps,** !ISC [cf. *Hall & Harbour 144*]
Physalis lobata: **Quincula lobata,** !ISC [cf. *Hall & Harbour 466*]
Pinus edulis, !ISC, "Pike's Peak" [cf. *Hall & Harbour 532*]
Pleurogyne rotata: **Lomatogonium rotatum** subsp. **tenuifolium,** !ISC [cf. *Hall & Harbour 477*]
Purshia tridentata, !ISC
Ranunculus alismifolius var. **montanus,** !ISC [cf. *Hall & Harbour 19*]
Rubus: **R. idaeus** subsp. **melanolasius,** !ISC
Saxifraga caespitosa: **Muscaria delicatula,** !ISC
Senecio: **S. hydrophilus,** !ISC, "Middle Park"
Specularia leptocarpa: **Triodanis leptocarpa,** !ISC [This specimen must have been collected on the eastern plains near the foothills of the Front Range, possibly near Colorado Springs.]
Thalictrum fendleri, !ISC [cf. *Hall & Harbour 8*]
T. sparsiflorum, !ISC [cf. *Hall & Harbour 9*]

Trifolium dasyphyllum, !ISC [cf. *Hall & Harbour 97*]
T. nanum, !ISC, "Pike's Peak" [cf. *Hall & Harbour 97*]
Woodsia scopulina, !ISC [cf. *Hall & Harbour 690* in part]

Expedition of 1864 4

In June 1864, Parry made a trip through Boulder County and the southern edge of Larimer County; in late July he went up the Clear Creek Valley, crossing into Middle Park to Hot Sulphur Springs in Grand County. Arps and Kingery (1972) state that "Parry joined William Byers (then editor of the *Rocky Mountain News*), George Nichols, and Jacob W. Velie in an attempt to climb Long's Peak. Velie, 1829–1908, was a zoologist and one time curator of the Chicago Natural History Museum." Following is Parry's narrative (C. C. Parry 1867). Two paragraphs dealing with Mr. Estes are taken from Byers's article in the *Daily News* (Denver, CO, Sept. 22, 1864). Parry's account of the Long's Peak ascent has been studied by David Hill, who found some confusion as to directions. Hill's corrections and elucidations are given in square brackets.

Wishing to ascertain more fully the peculiar features of the country lying north of this district [Pike's Peak to Clear Creek], I was induced to revisit this section a third time, during the past season, having Dr. J. W. Velie of Rock Island for my associate, who was especially devoted to Zoölogy [see Ewan and Ewan 1981, p. 228].

Leaving the settlements in May, we experienced the usual vicissitudes of climate on this exposed upland, in occasional storms of wind and rain. Coming in sight of the Rocky Mountains, we had a still stronger intimation of the severity of storms that had burst over the mountain slopes, by a very unusual rise in the Platte River, which in many places overflowed the wide bottom land, driving settlers to take refuge on the adjoining bluffs. Arriving at Denver City on the 2nd of June, the ravages of the severe flood of Cherry Creek, on the 19th of May, were still apparent, though the subsiding waters were then flowing barely six inches deep over a wide bed of sand. Still subsequent to this, deluges of rain visited the mountain region towards the sources of the Platte, flooding large sections of bottom land along its lower course, changing the beds of creeks and sweeping away a large portion of the then growing crops. This unusual and unexpected occurrence, in which heavy rains and melting snows combined to swell all the mountain streams, proved especially disastrous to the roads traversing this district, bridges being swept away, and roads and embankments washed and gullied to such an extent as to render the usual avenues of travel impassable. Owing to this condition of things, unusual difficulties and delays were experienced in making the proposed mountain explorations. These facts are still further worthy of note as illustrating the peculiarities of climate pertaining to this entire mountain region.

The first point of attraction was an elevated district, known to abound with small lakes, lying near the upper waters of Left Hand Creek, one of the tributaries of the St. Vrains. Our route lay through the excellent farming district of Boulder Creek, and passed over a series of uplands adjoining the foot of the mountains. Owing to prolonged spring rains, an unwonted luxuriance of vegetation was spread over the entire district, including the most arid sandy tracts. The streams crossed on our route were high and in many instances difficult of fording. The mountains were entered by a narrow cañon [Boulder Canyon] two miles south of Boulder City. [Actually, Boulder Creek flows east through the city of Boulder! The canyon south of Boulder would be Eldorado Canyon of South Boulder Creek. However, the party evidently did enter Boulder Canyon.] By a series of winding and steep ascents, this cañon penetrated the first range of mountains, which were composed of metamorphic [arkosic conglomerate] rock [the Flatirons] inclined at a very sharp angle with the horizon. Succeeding this is the usual form of coarse-grained felspathic granite, traversed at various points by veins of milky quartz. Thence by a series of moderate elevations, passing over rocky knolls or winding through level uplands agreeably diversified by pine groves and grassy swells, we reach the foot of Gold Hill, having thus in a distance of about seven miles attained an elevation of 1,500 feet above the base of the mountains. To reach this higher elevation we are forced to make a still more abrupt ascent by numerous zigzags till we gain the commanding height occupied by the mining settlement of Gold Hill, being 8,636 feet above the sea. Following thence the divide between Four-mile Creek (a branch of North Boulder) and Left Hand Creek on our right, we come to a chain of small lakes mainly occupied by rushes and aquatic plants, emptying by several small outlets and by a steep descent into Left Hand Creek.

At this point, on the borders of one of these lakes known as Osborn's Lake [according to Ewan (1950), this is one of three small ponds about 3 miles west of Gold Hill on the railroad from Boulder to Ward; Parry collected the type of *Nuphar polysepalum* here], we made a stationary camp—a series of barometric observations showing its elevation to be 8,821 feet above the sea and 3,300 feet above the base of the mountains, two miles distant. The region immediately adjoining is composed of mountain swells alternating with depressions and open valleys, which latter, when assuming a basin shape, are occupied by small lakes; the ridges are mainly occupied by a growth of *Pinus ponderosa,* mostly of small size. The larger streams (such as Left Hand Creek) lie deep below the general level and are reached by very steep descents. From elevated points there is a fine view of the Snowy Range, the main ridge being about ten miles distant. Farther north, Long's Peak is conspicuous, plainly exhibiting on its southeast slope the rugged features which subsequent explorations fully verified.

Among other conspicuous points, our attention was specially directed to a smooth, rounded peak, apparently of easy ascent, at that time pretty uniformly covered with snow, but showing several bare spots on its more abrupt sides. After taking as thorough an observation as we could obtain from different points of view of the intervening country, we concluded to attempt the ascent as early in the season as the 14th of June. In order to devote as much of daylight as possible to the severe labor of accomplishing the ascent, which would need to be comprised within a single day's journey, we started with a pack animal on the afternoon of June 13th and camped at the head of a small grassy valley where the heavy timber growth commences, having an elevation of 9,346 feet.

On the following morning we started at daylight, passing nearly due west over ridge and vale, through deep pine woods and intervening swells, encountering here and there masses of fallen timber, and again threading our way through dreary stretches of burned woods, rarely getting a near view of the Snowy Range which we were aiming to reach. Soon, occasional patches of snow lay along our path, rapidly melting under the rays of a summer sun; then, still farther on, continuous snow banks in which the alpine forests were deeply imbedded, occupied our path, forcing us to make short turns to avoid heavy drifts, and occasionally requiring fatiguing efforts to extricate ourselves from sudden plunges into treacherous holes. Meanwhile our route was somewhat enlivened by the discovery of beautiful, clear alpine lakes bordered by sheets of ice, and reflecting from the clear waters the somber forests of spruce by which they were surrounded. On one of these lakes, to the great delight of my zoölogical companion, an arctic loon was quietly sleeping. Still working our way westward, we encountered torrents of water, derived from melted snow, dashing and foaming over icy beds;

across these we were forced to pick our way, glad to avail ourselves of a fallen tree or slippery snag, to save a plunge into its chilly waters. Farther on, the unmistakable characteristic of an approach to that marked line which terminates the growth of trees, is apparent; the country opens up; the bare snowy range stands before us in bold outline, and then commences the steep slope that leads directly to the snowy crest. Our observations here show a timber line considerably lower than that noted farther south, being here 11,325 feet above the sea level.

From this point, a steady and continuous climb calls out all our nervous and muscular energies. The wintry summer has not yet had time to unfold her floral beauties, and animal life is scantily represented by the Rocky Mountain ptarmigan, now just exchanging its garb of winter white for the mottled colors of summer. Occasionally, from an area of rough rocks, the mountain badger [marmot] shuffles over sheets of snow to reach some more secure retreat, while here and there the diminutive alpine hare [pica] utters its sharp bark. Slowly and steadily we move on toward the highest point, which almost seems to recede from our advance; but finally the last steep ascent is gained, and we stand on the narrow crest overlooking a vast scope of country.

Middle Park, with its irregular undulations and open valleys, is seen to the west, and the sea-like expanse of the Plains is visible to the east; just below us, on the northwest slope, lies a deep gorge, in which the smoothly scooped bottom and polished rocky sides plainly marked the bed of an ancient glacier. Through this chasm, which is seen to penetrate deeply into the mountain mass, flow some of the extreme sources of the St. Vrains, a sharp and abrupt divide separating it from a similar gorge on the western side, which leads towards the main valley of the Grand in Middle Park. By this route it seemed practicable by a short tunnel to connect the two mountain slopes and, should the elevation on either side be found accessible by easy grades from the base of the mountain, the important problem of a direct railroad pass through the principal range of the Rocky Mountains would here find a practicable solution. From this elevation a number of lakes were brought distinctly to view, scattered at different points on the eastern slope of the mountains, while, ten or twelve miles distant, in a northwest direction, towers the rugged form of Long's Peak. The barometer was set up on the highest point of this snowy crest by means of a temporary tripod composed of the different parts of the portable case, and a series of observations were made with the following mean result: barometer, 18.502; thermometer attached, 42°F.; thermometer detached, 35°F. This gives an elevation (according to the computation of Dr. Engelmann) of 13,402 feet above the sea level. We supposed, at the time, that we were on a peak which is laid down on recent maps as Mt. Edmonds, but finding afterwards that we were mistaken in the point thus designated, at the suggestion of my companion, Dr. J. W. Velie, we concluded to affix to this well-marked elevation the name of the distinguished naturalist Audubon.

Having thus completed the necessary observations, we commenced the descent at 3 P.M. and, varying our route somewhat by selecting a more direct and abrupt slope, by a series of slips and plunges far from agreeable, we were overtaken by night before reaching the farthest limits of snow, and we arranged our night bivouac on the lee side of a precipitous rock. Here, before a blazing fire, we made an excellent supper of fresh ptarmigan, and enjoyed a somewhat repose from the labors of the day. Returning from this excursion to our permanent camp near Osborn's Lake, a few additional days were spent in exploring the adjoining country and in securing specimens of its peculiar plants and animals. The pond lily of these alpine lakes was only yet in bud, but the characters were sufficiently apparent to determine it to be an undescribed species—to which, from its most characteristic feature, I have given the name *Nuphar pictum* [this was officially named *N. polysepalum* Engelmann]. A visit to this same locality in the month of August enabled me to collect full material for its description. It is apparently quite local in its distribution, as amid a great number of lakes, lying along our route, it was only met with in two, adjoining our stationary camp. Doubtless it will yet be found elsewhere in similar locations, but still it must be regarded as a rare plant, well worthy to be rescued from its obscurity, and from possible future extinction, by horticultural enterprise.

Returning to Boulder City in the latter part of June, we again essayed to penetrate the mountains by the ordinary route of travel up Clear Creek. Reaching Empire City in July, we again established a permanent camp in this delightful summer retreat, revisiting the various alpine peaks, over ground rendered familiar by two previous seasons of exploration.

Hot Sulphur Springs

In the latter part of July, the mountain passes having by that time become passable for pack animals, we made an excursion to the *Hot Sulphur Springs* of Middle Park, passing over the Vasquez trail on the route selected for a wagon road, which had been worked to near the foot of the range, on the east side, in the summer of 1863. The divide here (being a little more elevated than Berthoud's Pass) reaches just to the limit of timber growth and, by a smooth open swell, passing on either hand steep rocky slopes, leads down on the western slope to the headwaters of Dennis Creek. The valley along this route, being more open than the corresponding one by way of Berthoud's Pass, is better adapted for a wagon road, and the slope, being more prolonged, has an easier grade. Like the former, however, the lower part of the valley is much obstructed by beaver dams, though the greater width offers better facilities for road construction. The open Park is reached at nearly the same point as by the Berthoud trail, and thence by a succession of open meadows and sage plains, passing over various intervening ridges, we arrive at the main valley of the Grand River. At the foot of a ledge of fixed rocks, where Grand River makes a sweep to enter a deep cañon [Byers Canyon], a surface of white rock is exposed, at the upper part of which, on a cool summer morning, a light cloud of vapory steam is seen to arise. This light-colored rock is formed by the gradually accumulated deposit of the Hot Springs which, issuing from several orifices, formerly trickled over the rocky surface, leaving a white calcareous incrustation. Part of this issue has since been directed in a small stream, thence falling by a cascade of about ten feet into a natural basin; here it drops onto a smooth, pebbly surface, sending up a shower of glittering spray and, passing through an arched cave, it reaches the waters of Grand River by a subterranean passage. The temperature of the water, at its several issues, is 112°F., and at the point where it falls into the basin, 110°F. A smaller spring, less exposed to the external air, shows a temperature as high as 115°F. The water is clear, has an agreeable softness to the touch, and a distinct saline odor. Along the course of its several streams floats a slimy growth of feathery *Confervae* [green algae], exhibiting a great variety of colors, including pure white, red, green, and blue. Animal life is also exhibited in these tepid waters, in a small mollusc of the genus *Limnaea* [*L. caperata* and *L. elodes*]. The sulphurous odor is not very perceptible, and to the taste, though disagreeably tepid, there is sufficient sparkle to render it not unpleasant.

The bathing arrangements are of the most primitive character. You stand under the falling stream as you are able to bear it, and allow it to fall with a dash on different parts of your body, till its diffusive warmth penetrates your system. The first impression is that of scalding, and you naturally spring away with a scream; by repeated trials, however, it becomes first supportable, and then delightfully agreeable. When not too long indulged in, its effects are both refreshing and soothing; but a prolonged stay is apt to beget lassitude. If a sweating process is desired, you wrap yourself up in a saturated blanket and lie down in an adjoining cave, which keeps a uniform temperature of about 90°F.; here perspiration starts from every pore till you are glad to emerge and submit to the final process of brisk rubbing. There is no doubt that these springs possess valuable medical properties which will be in due time appreciated; they are even now resorted to with benefit in cases of rheumatism, cutaneous eruptions, and other general disorders of the system arising from exposure, sedentary confinement, or intemperance. Doubtless no little of the benefits thus derived are due to the necessary exposure to out-of-door life, the excitement of travel and changes of scene, as well as the too often neglected means of personal cleanliness; but hereafter, when experience and investigation shall have determined the actual value of these waters as a medical agent, this as well as the agreeable accessories connected with romantic scenery, clear atmosphere, and the various active amusements of hunting and fishing, will render this a place of fashionable resort.

The character of the geological formation, as exhibited through this entire section of Middle Park, furnishes a satisfactory explanation of the phenomena exhibited in these hot springs, of which many others will doubtless be discovered in different parts of this section of country. The numerous cañons of Grand River bring to view a great variety of geological sections, showing both granitic protrusions and highly inclined stratified rocks in different states of metamorphism. The adjoining hills and mountains are frequently capped with basaltic or vesicular trap, all showing a previous active condition of volcanic phenomena. The general surface of the country is thus rendered irregular and varied, and the existence of hot springs represents the present condition of the active volcanic forces, now nearly extinct. The usual variety of volcanic products is further exhibited in the different qualities of soil, being in some portions coarse, arenaceous, and barren, while others exhibit a remarkable fertility, which is only checked by the severity of the seasons, in which killing frosts occur during every month of the year.

At lower elevations, however, more remote from the high Snowy Mountains, the valleys which can be irrigated are doubtless well adapted to agriculture; and in many other portions the uplands furnish an unlimited range for pasture. The character of the seasons is however quite variable, with occasional severe winters and heavy snows, burying up large tracts of country; while the summer rains, owing to the irregular distribution of high mountains, are uncertain and widely diverse in their continuance and in actual amount of precipitation. Still there is doubtless a large scope of habitable country which, in the development of its true natural resources, will be able to support a permanent population whenever it shall be made accessible by ordinary means of conveyance, especially should it eventually be found to lie along the main trunk of the great Continental Railway.

Excursion to Long's Peak

By the middle of August the snow had so far disappeared from the higher peaks and ridges, and the heretofore swollen streams had so far contracted, that we concluded to attempt the ascent of Long's Peak. Our route in this direction took us again to the eastern base of the mountains; thence passing over the uplands adjoining and crossing the numerous valleys and streams which, fan-shaped towards the mountains, converge in their easterly course to form the main stream of the St. Vrains. The cultivated crops in all these bottoms, where they had not been washed away by spring floods, exhibited unwonted luxuriance, most of the small grains being already harvested. Occasional *buttes* of basalt [North and South Table Mountains, west of Denver] formed prominent landmarks along our route, while at other points horizontal ledges of light-colored sandstone stretched in irregular mural bluffs along the borders of the principal streams [Laramie–Fox Hills formation south of Boulder]. Associated with this loosely coherent sandstone we met with various seams of lignite, in some localities attaining a thickness of five to eight feet. Where thus developed, these beds have been mined to some extent for fuel, the product finding a ready market at Denver City, thirty miles distant.

Passing thence towards a low break in the mountains where the main stream of the St. Vrains debouches on the upland plains [the present Lyons], we follow up a somewhat broad valley, bounded by mural ridges [Steamboat Mountain] of metamorphic [sedimentary Lyons sandstone!] rock showing a strong dip to the eastward. The various exposures here brought to view exhibit a series of perpendicular escarpments to the west, while its eastern face presents a prolonged irregular slope. The valley here shows a very fertile soil, easily irrigated, but subject to occasional floods. The low bottoms were occupied by a dense growth of annuals, consisting principally of *Helianthus,* intermixed with a perfect maze of willows, intertwined and bound together by spreading vines of *Clematis* [*ligusticifolia*]. These thickets were scatteringly occupied by clumps of wild cherry (*Prunus* [= *Padus*] *virginiana*), at this season crowded with profuse clusters of its purple fruit. Besides this, we find in abundance and of excellent quality the ripe fruit of *Ribes aureum,* exhibiting several distinct varieties including deep purple, light amber, and dull red [probably including *R. cereum* and *R. inerme*]. To each of these varieties there is a very distinct flavor, combining an agreeable mild acid with the peculiar odor of wild currant; and the berries are often fully half an inch in diameter and three-fourths of an inch in length. It is somewhat remarkable that this species, of which the cultivated product is considered worthless, should here in its wild state possess such desirable properties as a native fruit.

Proceeding onwards, following up the most northern branch valley and crossing by frequent fords the rapid stream, bedded with rounded pebbles, we are forced by precipitous rocks, hemming in the valley on either side, to the adjoining hills; and thence, by a very rugged route, making a somewhat rapid ascent, the stream terminates near its head sources in an open basin valley, quite picturesque in its surroundings, luxuriantly bedded with nutritious grasses and agreeably set off on either hand by pine-clad mountains. Thence, by an easy divide, we pass down into the valley of Big Thompson Creek, where the valley spreads out in broad undulating swells to form Estees, Park. Through this beautiful upland park the main creek pursues its meandering course, its tributary waters flowing direct from the higher points of the Snowy Range, a short distance below entering a deep cañon to pursue its intricate course through the intervening mountains to the great Plains.

Mr. Estes (Parry always used the spelling Estees), while hunting in the summer of '60, found this park. It was alive with game, and that fact, together with its beauty and luxuriant growth of grass, induced him to locate. He built a cabin, and a year or two after, took his family there to live, but they are getting tired of the solitude and we suspect would like a change. The picturesque will do for a time but, like everything else it grows monotonous. Eventually this park will become a favorite pleasure resort. Probably by another season the road will be so improved that a carriage can go from Denver directly to the foot of the Snowy Range, and the drive will be but a day and a half to reach such magnificent prospects and surroundings as the imagination can hardly paint.

Mr. Estes turned his attention to stock raising. He has a considerable herd of cattle, a few horses and mules, and every animal is as fat as it can roll. He makes butter and cheese for his own use, but not much more because of the difficulty of marketing it. His stock never leaves the park because they have no reason to. Practically it is fenced by a lofty stone wall, for there are but few places where animals could pass out if they would. Their range is over some thousands of acres and they go and come as they please. Any quantity of hay can be cut, and some is being saved this year, though hitherto cattle have wintered and done well without feed except what they find for themselves. Though considerable snow falls the wind blows it off many parts of the park so that an abundance of grass is always bare. In the fall, immense herds of elk, deer, and other game come into the park and from them they obtain their meat. The creek abounds with the most delicious trout, and in their seasons the valleys produce abundance of wild fruit so that these pioneers enjoy an actual Arcadian life.

We had found in our day's travel an abundance of mountain grouse, and the young ones were of just the right age for the frying pan. A dozen or more of them had fallen at the bidding of our ornithological sportsman, Dr. Velie, who is a capital shot, and besides dining on "spring chicken" upon the verdant banks of the Little Thompson, we had brought with us a not very meager supply of the same luxury. We enjoyed a hearty supper, to which was added some freshly caught trout, and prepared for a climb towards the peak the following day.

Not once during the day's travel had we caught sight of the great mountain. An intervening wooded range upon our left had shut it out from view, but as we turned the point and came upon the house of Mr. Estees, the object of our visit was again before us, directly up and seemingly at the head of a little grassy valley, one arm of the park itself. It seems very near and in the setting sun its snowy helmet looked cold and cheerless. We spent an hour in scanning its massive sides and speculating upon various routes by which to reach its summit, and settled upon but one thing. That was to try the ascent from a new point. All who had ever started up it had gone from this point up the little valley before spoken of and then up the eastern slope of the mountain, and they had invariably returned unsuccessful, pronouncing the ascent of the highest summit impossible. A careful examination with a glass convinced us that they were correct so far as their means of observation extended, but appearances indicated that the ascent might be made from the northwest. In consequence we determined upon that course.

At this point is the last and only settlement on our route, being about twenty miles, by the trail followed, from the foot of the mountains. Long's Peak is here plainly visible in all its rugged outlines, the highest summit bearing SSW (magnetic) probably ten miles distant in a direct line. A series of lower peaks on the east connects [Mount

Meeker] with the culminating point by a sharp ridge gashed by perpendicular chasms [the Notch]. A sheer precipice, partly hid from view by intervening mountains, forms the northern [eastern!] face [the Diamond], while on the west it connects with the main range by an irregular crest, forming a jagged descending ridge [Keyboard of the Winds]. With these unfavorable prospects for making the ascent of the main peak, we decided to vary from the route heretofore taken, which followed up a more gradual eastern slope to the lower range of peaks, and our course was directed toward the low, ragged ridges lying on the western slope, in hopes that some of these might afford a foothills to scale the highest tabled summit; at least we could reach its very base and get a fairer estimate of its actual height above us as than from a distant view.

To reach the actual timberline with pack animals and establish a near camp from which our examination could be made of the main steep ridge made up the first day's journey. We left the main valley of Thompson Creek, which bears more directly west, and crossing two considerable divides, came on a small mountain stream [Boulder Brook] leading direct towards the main peak. To follow this up through the obstruction of fallen timber, avoiding precipitous rocks, or working our way through tangled mazes of wooded thickets, required no small outlay of persistent and laborious effort. Catching occasional glimpses, through the open timber growth, of the high peak, whose near view showed its distinct outline, was encouraging; pausing meanwhile to gather the luscious berries of the mountain whortleberry [*Vaccinium myrtillus*], or to take a refreshing drink from the clear mossy brook that flowed in our direct course, we finally reached that sharp line where timber growth ceases, and established our camp at its upper edge, At this point we found needful shelter and firewood for ourselves, and a patch of alpine grass for our animals. A series of barometric observations shows here an elevation of 10,800 feet above the sea, with which to commence our climbing on the succeeding day [timberline on Boulder Brook is 10,800 ft. according to Long's Peak Quadrangle].

Such elevations are not often attained in this country, at least for camping purposes, and the interval between daylight and dark was variously employed in investigating its peculiarities. Some of the more adventurous and eager ones made their way up the bare slope of rocks to the base of the main peak, to return at dusk with a discouraging account of its inaccessible character; others found amusement in noting the peculiarities of vegetation at this exposed locality where, within a short distance, trees dwindle down to shrubs or bend their matted foliage so as to make up in horizontal extension for their lack of height.

There were rare alpine plants growing in exposed places, and various mosses and lichens helped to add an Arctic character to the bald scenery.

The evening was devoted to a discussion of routes and probabilities for the following day, before a blazing campfire; the night was rendered brilliant by twinkling stars in the serene atmosphere; and anon the advent of a nearly full moon cast spectral shadows through the somber and blighted forest around us; and then followed a refreshing rest, undisturbed by dyspeptic nightmares or outside intrusions, till the breaking morn routed us from slumber to prepare for the labors of the day.

Disposing of all unnecessary encumbrances, though protected by ample clothing and taking a pocket lunch, we commenced the first steep ascent. Extricating ourselves from the dense mass of bushy growth which lies above the timberline, we encounter the usual bare alpine exposures. The rocks, variously shaped by attrition and exposure to atmospheric agencies, lie either in loose fragments or imbedded in a coarse granite sand. In favorable situations vegetation acquires a foothold, forming dense patches of alpine sward, in the midst of which flourish the choice and variously tinted plants peculiar to this region. Trickling brooks, having their sources in banks of snow, form ribands of verdure, contrasting pleasantly with the dull-colored rocky ridges down which they course. These again at other points expand into mossy morasses, supporting a rank growth of sedges and other water-loving plants. Passing amid such scenery and gaining by steady climbing an elevation of 1,500 feet or more above the timberline we come upon a rock-paved tableland [the Boulder Field] where massive blocks strew the surface, varied here and there by patches of snow. This brings us to the base of the main peak still towering above us a thousand feet or more, but now plainly exposed in all its details of shape and position. On the side towards us, being the western [northern!] face of the peak, there is a steep slope, partially faced with snow, terminating higher up in a perpendicular ascent, and then another steep slope reaching to the flattened summit. Its northern [eastern] face shows a sheer smooth precipice of grayish colored rock extending from the very summit to a bewildering depth of probably 3,000 feet [the Diamond], terminating in a rocky gorge where snow-fed lakes [Chasm Lake, Peacock Pool] supply the headwaters of the northern branch of the St. Vrains. Two of the party ventured down this gorge, clambering along the rough *talus* at the base of the precipitous peak, aiming by this difficult route of descent and ascent to reach the easternmost peak [Mount Meeker]. In accomplishing this, they attained an elevation still considerably below the main summit, which was found inaccessible from this direction. Others, winding round the western face toward the south [north!], came upon the jagged ridge [Keyboard of the Winds?] that connects Long's Peak with the main range. On one of the highest points of this ridge, named (after one of the party) Velie's Peak [now Storm Peak], the barometer was set up, and a series of observations taken, giving an elevation of 13,456 feet above the sea. The estimated height of Long's Peak above this observed station was about 600 feet, which would give to this culminating point an elevation of 14,056 feet, which is thus seen to be somewhat lower than the more accurately determined height of Pike's Peak, which by barometric measurement in 1862 was found to have an elevation of 14,216 feet above the sea level.

Having thus tested to our satisfaction the inaccessible character of the highest tabled summit of Long's Peak, we made the best use we could, under the circumstances, of the elevated position we were enabled to attain. Below us, on the south, lay an open rocky gorge where, in irregular basins at various elevations, seven distinct lakes were counted; these go to form the head sources of the Big Thompson Creek, which thus hugs around the mountain mass on the north and west, thence meandering through deep valleys and open parks till it reaches the wide plains below. To the southwest, Middle Park, with its meadows and undulating ridges, was plainly visible. The divide which separates the latter from North Park could be traced only as a confused tangle of irregular mountains, North Park itself being hid from view by a succession of mountain ridges. Long's Peak is thus shown not to lie, as represented on most maps, in the actual divide or watershed of the Rocky Mountain Range, but, like Pike's Peak to the south, is an eastern offset, all its tributary waters flowing to the Atlantic slope.

Time is precious on these high elevations in noting the peculiarities of the vast region lying below and around us; but you are often in such places reminded still more forcibly of your elevated position by a chilly atmosphere which, coming in irregular gusts, strikes you with the force of a literal blow, from which you are fain to beat a retreat to more sheltered localities.

Consoling ourselves for our failure to reach the main summit by the reflection that we had done the best we could, and that others equally sanguine and still more adventurous could do no more, we turned our backs upon the towering summit, and by a rapid descent gained our previous night's camp by noon, and at nightfall reached the hospitable roof of Estees, to enjoy a supper of delicious mountain trout and the luxury of a soft bed.

One of the incidents of our return was the capture of a five-foot rattlesnake with thirteen rattles, which was caged in a coffee can with a cloth tied over the mouth, in which condition he became an occupant of the wagon in common with other curiosities in natural history.

The trip to Long's Peak and back can be made in five days, but it is better to take six, seven, or eight days for it. Ours occupied six and a half altogether.

List of Plants

6. *Salix chlorophylla:* **S. planifolia,** !ISC
106. **Viola biflora,** !ISC, "alpine ridges, June"
177. **Trifolium dasyphyllum,** !ISC
310. **Frasera speciosa,** !ISC
402. **Lycopodium annotinum,** !ISC [no date, but marked "Middle Park"]

Abies grandis: **A. bifolia,** !ISC
Allocarya scopulorum: **Plagiobothrys scouleri** subsp. **penicillata,** !ISC
Aquilegia coerulea, !ISC
Astragalus decumbens: **A. miser** subsp. **oblongifolius,** !ISC
Astragalus flexuosus, !ISC, "foot of the Rocky Mts., near Boulder City"
Astragalus tridactylicus, isotype, !ISC [The population occurs on Pierre shales about 6 mi. north of Boulder.]
Balsamorhiza sagittata, "Middle Park, July," !ISC [A superfluous label on the sheet is irrelevant: "Northwestern Wyoming Expedition, Capt. A. W. Jones, U. S. Engineers, commanding; No. 146, 1873."]
Eriogonum brevicaule, !ISC [very likely collected on the Pierre shales just north of Boulder]
E. pauciflorum: **E. exilifolium,** !ISC, "Middle Park" [The population near Granby is still the only locality known on the Western Slope.]
Festuca ovina var. *duriuscula:* **Vulpia octoflora,** !ISC, "covers large tracts of land adjoining Denver City"
Iliamna rivularis, !ISC
Jamesia americana, !ISC
Liatris scariosa: **L. ligulistylis,** !ISC, "Long's Peak" [The locality is probably Moraine Park, in Rocky Mountain National Park, where it is apparently extinct; it is a relatively rare plant in Colorado.]
Mammillaria: **Coryphantha vivipara,** !ISC
Muhlenbergia gracilis: **M. montana,** !ISC

Myosotis alpestris: **Myosotis asiatica,** !ISC, “Middle Park: Table Mt., July” [a very rare species known otherwise in Colorado only from the “Flat-Tops” in Garfield, Co. Table Mountain is in Grand County, Trail Mountain Quadrangle, and lies along the west edge of Lake Granby.]

Nuphar lutea [subsp. **polysepala**], isotype of *N. polysepalum*

Oenothera albicaulis: **O. pallida,** !ISC

Opuntia: Mixture, **Opuntia fragilis** and **Opuntia sp.,** !ISC

Oxytropis multiceps, !ISC

O. splendens, !ISC

Penstemon caespitosus, !ISC [the prostrate, hairy form of Middle Park]

P. fremontii var. *parryi,* isotype: = **P. watsonii,** !ISC, “Middle Park” [I annotated this incorrectly as *P. virens* Pennell.]

Phlox humilis, P. douglasii: **a. P. sibirica** subsp. **pulvinata; b. P. multiflora,** !ISC

Pinus aristata, !DUKE, ISC [probably collected at Silver Plume]

Polygonum polygaloides [subsp. **kelloggii**], !ISC, “Middle Park”

Pterospora andromedea, !ISC

Ranunculus adoneus, !ISC

R. eschscholtzii: **R. gelidus** subsp. **grayi** [Along Parry’s 1864 route, this is known only from the highest altitudes on Gray’s Peak. Parry could have collected this on Gray’s Peak in 1862 or 1872.]

Salix glauca var.: **S. brachycarpa,** !ISC

Saxifraga chrysantha: **Hirculus serpyllifolius** subsp. **chrysanthus,** !ISC

Sibbaldia procumbens, !ISC

Sidalcea candida, !ISC

Stellaria umbellata, !ISC

Streptanthus cordatus, !ISC [In Middle Park this species reaches its easternmost point and highest altitude along the Blue River south of Kremmling, along Parry’s route.]

Townsendia sericea: **T. leptotes,** isotype of *T. sericea* var. *leptotes,* !ISC, “Middle Park” [probably collected north of Kremmling, Grand County]

Vesicaria alpina; V. stenophylla: **Lesquerella alpina** subsp. **parvula** [probably collected in Middle Park north of Kremmling]

Viola palustris: **V. macloskeyi** subsp. **pallens,** !ISC

Wyethia amplexicaulis, !ISC, “Middle Park”

Observations on Snow Line and Timberline 5

Except for sporadic collections in Clear Creek Valley, the expedition of 1864 ended Parry's private field investigations in Colorado. Summing up some of his observations over the years 1861–1864, he presented a paper to the St. Louis Academy of Sciences entitled *On the Character of the Persistent Snow-Accumulations in the Rocky Mountains, Lat. 40°–41° North, and Certain Features Pertaining to the Alpine Flora* (C. C. Parry 1867).

A long-felt desire to become personally acquainted with the various forms and characteristic features of our Rocky Mountain Flora, having resulted in three different expeditions to that region, there has thus accumulated on my hands, as an incidental fruit of my observations, some general facts in reference to the natural aspects of that region, involving certain scientific conclusions apparently worthy of record.

Nearly all my previous information in reference to persistent snow accumulations in high mountain ranges having been derived from works descriptive of the European mountains, I was for a long time greatly puzzled to conform what I actually saw in the summer snows of the Rocky Mountains with what I had read in reference to the Alps. How far actual experience and observation have led me to modify these views will appear from a brief review of the results here arrived at.

My first ascent to the Snowy Range was made on the 14th of June, 1861, from the upper waters of South Clear Creek, my reliable guide on this occasion being a foaming ice-cold brook, appropriately called Mad Creek, which brawled past my rude cabin door, and which I felt confident, if resolutely followed, would lead me in the most direct course to its snowy sources. With an enthusiasm which, on after experience I learned to temper with more deliberation, I climbed, panting, up the steep rocky slopes, forcing my way through underbrush and clambering over fallen timber, only allowing myself to rest as some new floral form drew my attention, still intent on reaching the open and commanding summits beyond the pine growth, to which my eyes had been so often directed from the lower plains and valleys. Such interruptions, however, increased with every step as I drew near to the sharply defined line that limits timber growth, the straggling alpine plants increasing both in number and variety, till I was fairly bewildered with the strange novelties by which I was surrounded.

The first patches of snow encountered were the wasting remains of winter drifts, entangled in the deep woods which mark the upper belt of timber; these drifts, stained and spattered by the fallen leaves and decayed bark of overhanging pines, were rapidly melting in the warm atmosphere, the spongy bed of moss and decayed vegetation underneath being saturated with the icy waters that oozed from their sides. The snow itself, made up of small rounded grains, showed that it had been subjected to alternate thawing and freezing, by which the original feathery mass was converted into agglutinated ice. Some of the smaller trees bore evidence of the depth of snow in which they had been buried, by the abrupt inclination of their lower branches, forming a sort of circular tent wall around each trunk, which, as the melting snows gave way, left them in deep pit-holes, apparently sunk away from the general snow level.

At and beyond the timber line the snow lay in patches of greater or less extent, occupying depressions in the open valley, or smoothly filling up broad scooped recesses on the steeper mountain slopes. Along the immediate stream borders and frequently over-arching the rushing waters, the snow still lay deep, though unsafe to the foot, the constant wasting underneath weakening the support, and by abrupt plunges dropping the unwary traveller into treacherous holes. The warmth of the soil covered by the wasting snows was farther evidenced by the shooting forth of green leaves and occasional blossoms through a thin covering of snow. In these few observations was mainly comprised all the definite information derived from my first visit to the Snowy Range at its lower slopes.

Subsequently, however, as practice improved my climbing abilities, and I was enabled to scale the highest summits and look down on snow fields and ice-girt lakes, the question came up: Where is the snow line? At what determinate line do the regular winter snows increase beyond the power of the succeeding summer's sun to dissolve, thus leaving a constantly-accumulating mass to give origin to glaciers? In pursuing this investigation I climbed repeatedly to the highest summits within reach, attaining elevations of 13,000 to 14,000 feet above the sea. Still the same general features were observable, vegetation, indeed, becoming scant, but never entirely wanting, even at the highest elevations; no permanent snow deposits of which it could be said that they must necessarily increase from year to year. The fact of the largest bodies of snow being met with in depressions which, when filled up to a certain point, remained nearly stationary and did not accumulate by drifting more than the average heat of summer could dissolve—the entire absence of anything like glacier phenomena—soon satisfied my mind that the true "snow line" as understood in European countries was not reached, at least in this particular region.

Being thus fairly divested of all preconceived notions, I was in a proper frame of mind to apply the facts at hand to account for the phenomena to be explained. And now came up distinctly the problem, to explain the persistence of snow through the warm summer months in this mountain district, lying below the true snow line. Here, then, note first, an uneven surface, variously exposed to the direct action of the sun's rays, on which there is an average annual precipitation in the form of winter snow which, if regularly spread over the entire surface of the ground, would disappear in the warm season sufficiently early to allow the maturing of its ordinary vegetation; but owing to the surface irregularities it cannot be thus evenly spread, and furthermore, the light character of this alpine snow renders it peculiarly liable to be acted on by the fierce winds sweeping over these open wastes, driving into sheltered hollows and piling the wintry product in huge drifts, especially abundant in the upper valleys.

As a direct consequence of these conditions, we find snow accumulated in all the natural recesses, while the more exposed peaks and ridges are comparatively but thinly covered; hence, when the lengthening days and the more direct rays of the summer sun exert their melting power on these wastes of wintry snow, the thinner portions yield first to its influence, and as the protruding rocks and exposed darkened surfaces become heated, vegetation starts with unwonted vigor from its long winter sleep of nine months continuance; rivulets gush freely from the contracting edges of snow drifts, swelling the alpine brooks, and pouring their united tributes of melted snow water through the lower valleys to the distant plains.

In this condition of things there is of course no opportunity for the development of glacier phenomena; no accumulated body of snow above to exert pressure on the mass beneath, thus converting its névé to glacier ice. In fact the higher peaks and crests, being more exposed to the action of fierce winds, afford the scantiest space for the falling snow to rest upon, and are generally the first to become bare. Besides, having no source of supply but what is derived directly from the atmosphere, all that is borne down by the wind is so much taken from its power of resisting the continued action of the sun's rays. It is, indeed, possible that this transporting agency of wind may constitute the process by which, in this region and the more elevated districts of Mexico and the South American Andes, the culminating points are prevented from accumulating that local amount of winter snow which would otherwise be sufficient to establish a definite snow line, and thus give origin to that slower process of descent represented by the glacier phenomena of the Old World.

It is on the higher mountain slopes above the timber line that the evidences of ancient glacier action are most conspicuous, being here less obscured by the mingled product of subsequent surface denudation. Some of the upper valleys thus exhibit, in their deeply scooped beds and polished rocky sides, the traces of ice movements that have since given place to the ordinary abrading action of running water derived from melting snows.

At many other points along the irregular slopes of the highest crests, we meet with what are termed in the expressive language of mining prospectors as *sags,* representing as it were, slips, scooped out from the sides of the mountain ridges, and which may be properly regarded as incipient valleys, abandoned in their present unfinished state by their parent glaciers, and now only affording a bed in which the light-drifting snows of winter may find a resting place.

The outweathering of the natural rock exposures, as exhibited on the peaks and flattened summits of the highest elevations, is instructive as showing the combined effect of frost and atmospheric denudation. Nowhere do we meet with extensive smooth surfaces of level rock, such as we might suppose would result from uniform surface abrasion, but, instead of this, loose angular blocks of all sizes, piled in every conceivable form, either loosely, leaving extensive cavities, or imbedded in a coarse granitic sand. These detached blocks, variously spotted over with lichens, serve to give a peculiar aspect to the scenery, and the persistent irregular position can only be satisfactorily explained by referring to the action of melted snow penetrating the fissures which, by subsequent freezing, quarry out the immense blocks that strew the surface. These, once detached from their beds, are readily subjected to the displacing movements of underlying bodies of ice, melting unevenly according to differences of local exposure, and hence frequently poised in unstable equilibrium, rendering the footing insecure and at times dangerous.

Along the mountain front these features are of course modified by the action of running waters and occasional avalanches of snow and loosened rocks, all confusedly mixed up with the results of ancient glacier action. Hence it is not uncommon along the flanks of the Snowy Range to encounter extensive tracts completely bedded with fragments of rock of all sizes, and variously shaped by attrition. Over these, traveling can only be accomplished by a succession of leaps from one rock to another, while in the burrowing recesses and crevices the peculiar alpine plants and animals find needful shelter from the rigors of an arctic winter. Through these loosely aggregated beds the melted snow waters percolate and work their sinuous course by unseen channels to form the alpine brooks below. Occasionally a shallow basin, accumulating in its bed the finer sediment washed down from above, supports a rank growth of vegetation, being thus gradually converted into an alpine morass. At other points, smooth slopes are covered with a rich alpine sward composed of densely tufted plants with spreading fibrous roots and matted foliage, whose variously colored flowers complete an enameled carpet, attractive alike to the eye and foot of the mountain traveler.

These several enumerated points comprise the main features of scenery presented during the summer months on these bare exposures, wonderfully varied, it is true, and of which a mere verbal description can give but a very inadequate conception.

The Timber Line and the Upper Limit of the Growth of Shrubs

No feature of Rocky Mountain scenery is more strongly marked than that which determines the upper limit of the growth of trees. This can be readily traced by the eye from a distant view, embracing a wider scope of mountain exposure in which the slighter inequalities are smoothed down to a more uniform level, and thus presents a well marked horizontal line.

This feature is most distinctly marked when the green foliage of the upper pine growth is brought in contrast with a background of glittering snow, such as prevails during the period of winter storms, extending from early autumn to late in the succeeding spring. But when the summer sun lays bare the exposed rocks and smooth alpine slopes, the contrast becomes less marked, and the line itself can be less regularly traced by the eye. Barometric measurements show that this timber line, with slight local variations, marks a very uniform elevation, gradually diminishing with a more northern latitude. Thus, Pike's Peak on the south shows a timber line having an elevation of 12,000 feet above the sea level, while Long's Peak, nearly two degrees farther north, marks the tree limit at an elevation of 10,800 feet. Intermediate points, as Gray's Peak and Mount Flora, give a comparative mean of 11,700 feet for their respective timber lines.

It would be particularly interesting to know whether this timber line, offering so peculiar and well marked a feature of alpine exposures in different portions of the globe, is connected with a certain range of mean annual temperature and, if so, what is the actual mean thus indicated.

A more simple explanation of the phenomenon in question has been suggested by my observations in the Rocky Mountains, and I am inclined to refer the limitation of tree growth to a certain range of minimum temperature. The facts especially bearing on this point are these: In the first place, there is a very marked distinction in the timber line itself, everywhere easily recognized, and, as we may say, marking at each locality two distinct timber lines, the one of very uniform elevation, consisting of thrifty, often large trees, of upright growth, and which terminates sharply, without any manifest dwarfing or stunting of regular growth, though mainly confined to species of pine that higher up become dwarfed and deformed in their struggle with the elements; this first and lowest I would designate as the *true timber line.* But again, above this, and straggling irregularly up sheltered ravines and rocky slopes, is a class of depressed tree growth, singularly deformed, often spread out in dense mats on the ground, forming almost inextricable thickets. At other points, it presents itself to view with blighted tips and twisted, gnarled lower branches and prostrate trunks, creeping snakelike through rocky fissures—everything betokening a struggle for existence. The arborescent forms here represented belonging almost exclusively to *Pinus aristata* and *Abies* [*Picea*] *engelmannii,* both of which, as far as known, are peculiar to the Rocky Mountains. The actual elevation of this line being determined altogether by the local peculiarities of surface, as affording necessary shelter, is of course quite irregular and undulating, and may therefore be properly designated as the *false timber line.*

My explanation of the phenomena here so plainly exhibited is this: The *true timber line,* indicated by the natural growth of erect trees (and which terminates sharply at a regular line of elevation), marks a point of *minimum,* or extreme winter temperature, below which no phaenogamous vegetation [flowering plants] can survive atmospheric exposure; only two or three of the hardiest forms of tree growth reach this limit, and these, endowed as it were with special powers of endurance, leave all others behind to mark the barrier beyond which the rigid elements absolutely forbid all organized growth that is not protected by the thick covering of winter snow. Such trees as venture above this limit can only survive by submitting to the condition of a winter burial, by which their otherwise erect forms are bent down and twisted to the earth, while all ambitious branches reaching into the sunlight of this arctic winter are inevitably nipped and lose their vitality. The features of growth thus exhibited in their summer resurrection present a grouping of the most strange and weird forms of struggling existence that mark so peculiarly the alpine scenery. Some future artist will here find scope for his pencil in portraying a character of Rocky Mountain scenery witnessed only in its perfection by the alpine explorers—when once seen never to be forgotten.

It appears on this view that what we have above designated as the *false timber line* exhibits only a particular phase of the alpine flora, represented elsewhere by the growth of shrubs, whose persistence to still higher elevations is due to the same general cause of protection by wintry snow from the otherwise killing effect of winter temperature. That shrub growth is thus limited is evidenced by noticing that the highest elevation of this class of alpine growth is attained by taking advantage of the shelter of ravines, where snow accumulates the heaviest and most constantly. The shrubs that thus attain to the highest elevations belong to the genus *Salix,* whose general mode of growth, consisting of slender, flexible twigs, easily bent down to the earth, peculiarly fits them for complying with the conditions of winter burial. Among the three or four species here represented, *Salix reticulata* takes the palm as the highest climber, being found at an elevation of 12,000 feet above the sea, its prostrate form enabling it to secure the necessary shelter at exposed situations.

In order to give some general idea of the character of our Rocky Mountain alpine flora, I submit herewith a list, as complete as my present means of information furnish, of the alpine plants met with in the district embraced within my personal observation. In this list I confine the term "alpine" to such plants as are met with on the bald exposures above the timber line; by a (*) prefixed, I would indicate those species which are exclusively confined to such localities, while others not thus marked are met with at lower elevations. The subjoined localities, wherever given, denote that the species referred to is not peculiar

C. C. Parry (age 64, fide Ewan 1950). *Courtesy of Hunt Institute for Botanical Documentation, Carnegie Mellon University, Pittsburgh, PA.*

to the Rocky Mountains, but is also met with in the different regions there named, *Eu.* indicating Europe, and *As.* Asia.

This paper is significant in being the first one to deal with the problem of timberline in North America, an enigma that continues to the present time. It is also the first one attempting to sort out the plant geographical groups, especially the strong Asiatic element, represented in the Rocky Mountain flora. Of the 142 plants listed, 51, about a third, are recognized as also occurring in Asia. The list of species is omitted here because it has lost its relevance through later research.

Expedition of 1867 6

The 1867 expedition, for the Pacific Railroad Survey, followed the valley of the Huerfano River from near Walsenburg up to Sangre de Cristo Pass (close to La Veta Pass). It was on this expedition that Parry collected the type specimens of *Neoparrya lithophila* Mathias [= *Aletes lithophilus*]: "On rocks, Huerfano Mts., No. 83," erroneously thought to have come from northern New Mexico (Weber 1958). An extensive plant list was published (C. C. Parry 1870), including specimens from Kansas, Colorado, New Mexico, Arizona, and California, but the Colorado collection at Iowa State is quite small.

List of Plants

59. *Hedysarum mackenziei:* **H. boreale,** !ISC, "Sangre de Cristo Pass"
70. **Parnassia parviflora,** !ISC, "Huerfano, Aug. 1867" [This would be the upper Huerfano River at Sangre de Cristo Pass.]
88. *Seseli nuttallii pr. p.:* **Aletes lithophilus,** isotype of *Neoparrya lithophila,* !GH
104. *Linosyris depressa:* **Chrysothamnus depressa,** !ISC, "Sangre de Cristo" [If the locality data are correct, this was collected on Sangre de Cristo Pass, very close to La Veta Pass, on the headwaters of Huerfano Creek. Thus far we have specimens only from the Western Slope of Colorado, but the sagebrush meadows should be searched for the species. I suspect this specimen actually was the *Bigelovia vaseyi* collected in 1883 in the Gunnison Basin.]
115. **Bidens tenuisecta,** !ISC, "Fort Garland; Upper Rio Grande, Sept."
122. **Dyssodia papposa,** !ISC
148. **Penstemon albidus,** !ISC, "Smoky Hill Fork, June" [probably in southeast Kit Carson County]
149. **P. grandiflorus,** !ISC, "Smoky Hill, June" [now found on the eastern plains near the Nebraska border]
161. *Mertensia alpina:* **M. lanceolata,** !ISC
172. *Gentiana acuta:* **Gentianella acuta,** !ISC, "Sangre de Cristo [Pass]"
182. **Abronia fragrans,** !ISC, "Huerfano"

190. **Eriogonum cernuum,** !ISC, "Huerfano, Aug."
191. **E. effusum,** !ISC, "Huerfano, Aug."
194. **Goodyera oblongifolia,** !ISC
238. *Trisetum subspicatum:* **T. spicatum** subsp. **congdonii,** !ISC, "Sangre de Cristo"
243. *Notholaena fendleri:* **Argyrochosma fendleri,** !ISC
364. **Monroa squarrosa,** !ISC

Artemisia parryi: type collection, from "Huerfano Mts., N. Mex." [actually Sangre de Cristo Pass, Huerfano County] = **A. laciniata** subsp. **parryi,**

Asclepias brachystephana: **A. uncialis,** !ISC

Calylophus lavandulifolius, !ISC

Cheilanthes lanuginosa: **C. feei,** !ISC

Juncus nodosus, !ISC

Penstemon cobaea, !ISC, "Smoky Hill, June" [either in Kit Carson County or in adjacent Kansas. This was collected in 1867, not 1889!]

Sorghum nutans: **Sorghastrum avenaceum,** !ISC

Woodsia oregana, !ISC

Dedication of Gray's and Torrey's Peaks, 1872 7

In the summer of 1872, Parry, Asa Gray, E. L. Greene, and others ascended Gray's Peak, for the dedication of Torrey's and Gray's Peak (see Weber 1972). Torrey was infirm and did not come. This occasion must have been a very great source of pride and accomplishment to Parry. An account of this celebration was given in *The Georgetown Miner,* 6(14), Aug. 22, 1872.

THE GRAY'S PEAK PARTY
A PARTY OF TWENTY-ONE WITH ASA GRAY AND HIS WIFE
INTERESTING PROCEEDINGS ON THE SUMMIT OF GRAY'S PEAK

A party of Ladies and Gentlemen visited Gray's Peak from Georgetown on the 14th of August. The day was calm, clear and beautiful. Some of the party, in order to lighten the toil of the journey, left Georgetown on the afternoon of the 13th, and the others joined them early in the morning at Bakerville. The party consisted of the following persons: Prof. Asa Gray, and wife, Cambridge, Mass.; C. C. Parry, St. Louis, Mo.; Rev. [E. L.] Greene, Greeley, Col.; J. H. McMurdy and wife, Georgetown; Dr. Munsom, Empire; D. S. True, S. F. Smith and wife, C. E. Putnam, J. D. Putnam, Davenport, Iowa; E. W. Thayer and wife, J. H. Thayer, Georgetown; Miss Kaiser, Davenport, Iowa; R. S. King, New York; R. Martin and wife, Rev. John Cree, Rev. Prof. Weiser, Georgetown.

This highly intelligent party reached the summit of Gray's Peak about 10 a.m., after pretty hard climbing. After looking around for some time, it having been agreed upon without the knowledge of Prof. Gray, to hold a meeting on the top of the peak, to give him a cordial welcome.

The meeting was called to order by J. H. McMurdy. Rev. J. Cree, one of the oldest citizens of Colorado, was called to preside. R. L. Martin, of Georgetown, was appointed Secretary.

J. H. McMurdy made a few very appropriate remarks in reference to the peak and its position. He stated that it divided the waters of the Pacific from those of the Atlantic, that this famous mountain had been called Gray's Peak in 1862, by Prof. Parry, and the peak on the north had, at the same time, by the same person been called Torrey's Peak, and that attempts had been made recently to change the names of those peaks without any good reasons, and he hoped this would not be done. He stated that our Army and Navy were represented, and also our explorers and our clergymen in Pike's Peak, Sherman and McClellan mountains, and Beecher mountain, and that our scientific men should also be represented. He was, therefore in favor of Gray's Peak and Torrey's Peak. Mr. McMurdy then proposed that Prof. Weiser now welcome Prof. Gray to the summit of the peak, whereupon the following address was delivered to Prof. Gray amid the most profound attention.

> Honored and respected Sir: It gives me no little pleasure to extend to you this welcome to the summit and center of the Rocky Mountains. And although many of us who surround you on this lofty and sublime spot are strangers to you, you are not altogether a stranger to us. The instructive books you have written, and the fame you have honestly and justly acquired, have gone as far as the waves of our civilization have dashed their fertilizing spray!—Most of us learned to honor you when we were students. We, sir, well recollect the difficulty we had in committing then to us, the hard technical terms of your Botany to memory, but our old Latin tutor used to repeat the old saying, *"Memoria augetur excolendo,'* and we found in after years that he was right.
>
> The human mind is a wonderful contrivance; unlike all other things of capacity, the more you put into it the more it will hold. And in no department of science is this peculiarity of the mind better illustrated than in your favorite science of Botany, and in no individual case do we find a more illustrious example than in your own. The man who can name, and rank and file more than 100,000 plants, must have a better memory than ever was possessed by the King of Pontus.
>
> You commenced your scientific career as a Geologist, and Geology may well regret your course in abandoning its hidden treasures for the more beautiful and gorgeous science of flowers. At quite an

early period of your literary career you commenced the study of science that had for ages occupied the attention of many of the greatest intellects of the world. This field of science had been cultivated by such men as Solomon, Theophrastus, Dioscorides, the Elder Pliny, the great Arabian Avicenna in the days of Charlemagne, Brunfels, Fuchs, and Bock, of Germany, Dodonaeus, Gesner, De l'Obel, Clusius, Turner, Ray and Lindley, of England, Morison of Scotland, Tournefort, Rivinus and Linnaeus, the de Jussieus, Brogniart, Candolle, and many other distinguished savants—where it was thought no new conquests could be won. Here in this gleaming field where it was thought forty years ago nothing could be picked up but a few worthless pebbles, you, sir, have gathered diamonds of the purest water. You with your honored teacher and life-long friend Dr. Torrey, have perhaps done more to simplify and enrich the science of botany than any of your predecessors. And in saying this, we do not wish to undervalue the labors of those illustrious men who have gone before us. It was your good fortune to be born in the first decade of the nineteenth century, when all the sciences began to assume a more practical form. It required two hundred years in the English world of letters to introduce fully the inductive philosophy of Bacon into the sciences.

For your labors in systematizing and arranging the science of Botany upon the broad basis of common sense, your countrymen have honored you. Standing as you do with your illustrious co-laborer, Dr. Torrey, at the head of your department of the natural sciences, your name and memory deserves to be handed down to posterity with honor as one of the chief promoters of true science in your day and generation. And although it may be true as the Poet says: "When fame a loud trump has blown her noblest blast, though long the sound, the echo dies at last, and glory like the Phoenix 'mid her fires, exhales her odors, blazes, and expires?"

Yes, sir, there is an honorable, well earned fame the best men have a right to covet. It is that fame that grows out of the consciousness that we have acted well and nobly our part in the great drama of life. This fame, sir, not only posterity, but your contemporaries will most cheerfully award to you. It has been said by one of your distinguished contemporaries, Wm. Cullen Bryant, that he has written his name all over this whole continent, upon our lakes and rivers, and on our broad prairies. But, sir, your name is destined to occupy a loftier niche. Your friend, Prof. C. C. Parry, has inscribed your name upon this lofty peak, this wedge that pierces the clouds and divides the waters of this continent, and there we hope it will remain and grow brighter and brighter as long as the waves of the sea dash upon our Atlantic coast, or roll in gentle murmurs upon the golden strand of our quiet Pacific.

We have met here today to welcome you to this, perhaps the highest spot in the United States, and we rejoice that he who has named so many of our beautiful mountain flowers has come into our midst to see with his own eyes on what a magnificent scale nature carries on her work in Central Colorado.

We, your friends, would say to you who now stand here upon the very central pinnacle of our widely extended country, as the inscription on the monument of Sir Christopher Wren, at St. Paul's says, "*Si monumentum Dei videres, Circumspice.*" We bid you welcome to this magnificent peak. May you, and your excellent wife, when you leave these wild and rugged scenes for your classic home in Cambridge, remember the pleasant gatherings we had on Gray's Peak. And although there are but few now in this stirring and bustling region who have time to study our beautiful flowers, and those gorgeous flowers are permitted to "waste their sweetness o'er the desert air," yet we hope the time is not far distant when there will be many in Colorado who, being stimulated by your labors and success, will study your favorite science. The study of no one science measures so correctly the progress of true civilization and refinement as Botany. Savage and barbarous nations pay no attention to this beautiful and elevating science. Hence, sir, you have been instrumental in promoting the civilization of your country, and for this we honor you.

It was meet that this magnificent peak should be called after your name, and your friend and fellow laborer and pupil Prof. C. C. Parry, who now stands by your side, conferred an honor upon his country and science when, in 1862, he named this grand old mountain Gray's Peak. And we who stand around you here today, in the presence of Prof. Parry, in the name of our whole country, in the name of Colorado, and in the name of science, now solemnly confirm and rectify the name of this mountain. Gray's Peak let it be called to the end of time! Yonder peak a short distance north of where we now stand, was also named in 1862 by Prof. Parry, Torrey's Peak, in honor of your teacher and fellow laborer Prof. Torrey. And although it is a disputed point which of these stupendous mountains is the higher, yet the general impression now is that Torrey's Peak is from 60 to 100 feet higher, and if on exact measurement this should be the case, we know you will not regret it. You are no doubt willing that your illustrious teacher and life-long co-laborer should occupy a higher niche in the temple of fame.

This, it is true, was not necessary to establish or perpetuate the fame of Dr. Torrey or yourself, for by your genius and industry you have both acquired honor and fame as imperishable as marble or granite. With these few remarks, in the name of this intelligent company, I now welcome you and your wife to the summit of Gray's Peak.

In reply, Prof. Gray made a short and appropriate speech. Among other things he said that he was not accustomed to public speaking, and that he could not express what he felt in his heart, but thanked the company for the honor they had shown him, but seemed to be much more solicitous about the honor of his friend Dr. Torrey than about his own. And especially as an attempt has been made recently to change the name of Torrey's Peak, he hoped the name of Torrey would be retained. The Prof. further said that he was now on his way home from a long journey to California and Oregon, that he had visited all the most famous localities on the Pacific slope, among which he mentioned the far-famed Yo Semite valley, but nowhere had he seen any scenery equal in grandeur and sublimity to the view from this peak. He had also travelled in Europe, but saw nothing there that surpassed this view.

It was then proposed that the proceedings be published in the *Colorado Miner,* after which the National Air, *My Country, 'Tis of Thee, Sweet Land of Liberty* was sung by the Ladies—and what was interesting, a son of the author of that patriotic song was present and helped to sing it.

The party then enjoyed the grand scenery for an hour, and then left for Georgetown, where they all arrived safe and sound at 8 o'clock, well pleased with their visit to the peak, from whose summit is had perhaps the most extensive and magnificent view in the world.

John Cree, Pres't of Party
R. L. Martin, Sec'y.

List of Plants

Again, the collections lack locality information but may be interpreted on the basis of the respective species' known ecology. Almost all of the collections are unnumbered. Those with numbers may actually have been collected in an earlier year and bear the wrong printed label. The flora represented by the collection falls into the geographic scope of the Clear

Gray's (left) and Torrey's Peaks from Argentine Pass, 1984. *Photo by David Hill.*

Creek Valley, Gray's Peak, and (a few) the Boulder area.

1. **Carex canescens,** !ISC [The numbers 1–5 refer to the list in a shipment to Stephen T. Olney, who replied in November 1872. See letter mounted with no. 3.]
2. *C. pyrenaica.* **C. crandallii,** !ISC
3. *Erigeron glandulosus:* **E. vetensis,** "*Erigeron, n. sp.*, see Gray's notes," !ISC
3. *Carex sp. indet:* **C. nebrascensis,** !ISC
4. *C. foetida:* **C. vernacula,** !ISC
5. *C. leporina* (= *C. ovalis* Good), *fide* Olney: **C. phaeocephala,** !ISC
168. **Senecio fremontii** [subsp. **blitoides**], !ISC
215. **Pedicularis parryi,** !ISC
363. **Poa arctica,** !ISC
367. *Aira caespitosa* var. *arctica:* **Deschampsia cespitosa,** !ISC
422. *Antennaria alpina:* **A. umbrinella,** !ISC
442. *Carex filifolia,* alpine form, "not in last year's coll.": **C. elynoides,** !ISC

Androsace septentrionalis, !ISC
Antennaria dioica: **A. microphylla,** !ISC
Aquilegia vulgaris: **A. saximontana,** !ISC

Arceuthobium americanum, !ISC

A. robustum: **A. vaginatum** subsp. **cryptopodum,** !ISC

Arenaria fendleri: **Eremogone fendleri,** !ISC

A. arctica: **Lidia obtusiloba,** !ISC

Arnica mollis: **Arnica cordifolia,** "*Arnica mollis* Hook.? var. It is your No. 2 of 1862, a reduced form.—Gray"

A. angustifolia: **A. mollis,** !ISC

Aster salsuginosus: **Erigeron peregrinus** subsp. **callianthemus,** !ISC

Brickellia grandiflora, !ISC

Calochortus gunnisonii, !ISC

Campanula uniflora, !ISC

Chionophila jamesii, !ISC

Claytonia chamissonis: **Crunocallis chamissoi,** !ISC

Cymopterus alpinus: **Oreoxis alpina,** !ISC

Diaperia prolifera: **Evax prolifera,** !ISC, "Boulder Valley, June"

Draba crassifolia, !ISC

Echinospermum floribundum: **Hackelia floribunda,** !ISC, holotype of *Lappula subdecumbens*

Erigeron acre: **Trimorpha elongata,** !ISC

E. grandiflorum var. *elatius:* **E. elatior,** !ISC

E. leiomerus, !ISC

E. macranthum?: **E. subtrinervis,** !ISC

E. uniflorum: **E. simplex,** !ISC

Eritrichium jamesii: **Oreocarya suffruticosa,** !ISC

Gentiana affinis: **Pneumonanthe affinis,** !GH, ISC [The label "1872" may be incorrect, for the species does not grow in the area covered; the collection undoubtedly came from South Park and might be attributed to 1861 or 1862.]

Gentiana tenella: **Comastoma tenellum,** !ISC

Geum macrophyllum var. **perincisum,** !ISC

Gnaphalium strictum: **G. uliginosum,** !ISC

Heuchera bracteata, !ISC

Hieracium fendleri; Crepis ambigua: **Chlorocrepis fendleri,** !ISC

Hippuris vulgaris, !ISC

Juncus parryi, !ISC

Kalmia microphylla, !ISC

Linosyris parryi: **Chrysothamnus parryi,** !ISC

Lloydia serotina, !ISC

Mertensia paniculata: **M. lanceolata,** *sens. lat.* (*M. viridis*), !ISC

Mimulus luteus: **M. guttatus,** !ISC

Mitella pentandra, !ISC

Muhlenbergia andina, !ISC
Oenothera caespitosa, !ISC
Oxyria digyna, !ISC
Oxytropis campestris var. **gracilis,** !ISC
Pachystima myrsinites: **Paxistima myrsinites,** !ISC
Parnassia fimbriata, !ISC
Paronychia: **P. pulvinata,** !ISC
Pedicularis groenlandica, !ISC
P. procera, !ISC
P. scopulorum, !ISC
Penstemon procerus, !ISC
P. secundiflorus: **P. virgatus** subsp. **asa-grayi,** !ISC
P. secundiflorus f. *alpina:* **P. hallii,** !ISC
Pirus americana: **Sorbus scopulina,** !ISC
Pleurogyne rotata: **Lomatogonium rotatum** subsp. **tenuifolium,** !ISC [The label gives the year as 1874, but this is probably an error.]
Poa serotina: **P. nemoralis** subsp. **interior,** !ISC
Polemonium humile: **P. pulcherrimum** subsp. **delicatulum,** !ISC
Polygonum bistorta: **Bistorta bistortoides,** !ISC
P. douglasii, !ISC
Primula angustifolia, !ISC
Pyrola rotundifolia var. *uliginosa:* subsp. **asarifolia,** !ISC
Ranunculus affinis: **R. cardiophyllus,** !ISC
Sagina linnaei: **Tryphane rubella,** !ISC
Salvia reflexa, !ISC
Saxifraga adscendens: **Muscaria adscendens,** !ISC
S. bronchialis: **Ciliaria austromontana,** !ISC
S. flagellaris: **Hirculus platysepalus** subsp. **crandallii**
Shepherdia canadensis, !ISC
Silene acaulis [subsp. **subacaulescens**], !ISC
S. menziesii: **Anotites menziesii,** !ISC
Stellaria umbellata, !ISC [alpine ecotype, *S. weberi*]
Thalictrum fendleri, !ISC
Trifolium nanum, !ISC
Vaccinium myrtillus: **V. scoparium,** !ISC
Valeriana dioica: **V. capitata** subsp. **acutiloba,** !ISC

Collections of 1873–1889 8

1873

Penstemon cristatum: **P. auriberbis,** !ISC, "Twin Creek, June" [Twin Creek feeds Lake George, in Park and Teller Counties. The species is known from near Colorado Springs.]

1874

The only reference to 1874 in Ewan and Ewan (1981) concerns a trip to southwestern Utah. The specimens in the following list represent a potpourri of species that might have been collected (with the exception of *Astragalus convallarius*) on the Eastern Slope of Colorado in other years. The same comment might be made for the following years, for which Ewan and Ewan give no information.

Astragalus junceus: **A. convallarius,** !ISC
A. hypoglottis: **A. agrestis,** !ISC
Atriplex argentea, !ISC, "No. 10, East Colorado"
Botrychium lunaria, !ISC
Senecio crassulus: **S. wootonii,** !ISC
S. andinus: **S. serra,** !ISC
Sedum rhodanthum: **Clementsia rhodantha,** !ISC

1876

Arctostaphylos uva-ursi, Spreng., !ISC
Ceanothus fendleri, !ISC
Pinus contorta [subsp. **latifolia**], !ISC

1878

37. **Lobelia siphilitica,** !ISC, "South Boulder"
32a. **Castilleja sessiliflora,** !ISC, "Colorado Springs, May 9"
56. *Vicia americana:* **V. ludoviciana,** !ISC, "Colorado Springs"
60. *Synthyris plantaginea:* **Besseya plantaginea,** !ISC, "Manitou, May 14"
199. **Lupinus pusillus,** !ISC, "Denver"
211. *Epilobium coloratum:* **E. ciliatum,** !ISC
242. *Lathyrus palustris:* **L. leucanthus,** !ISC
320. *Senecio aureus:* **Packera werneriifolia,** !ISC, "Mt. Leavenworth" [Clear Creek County]
386. *Epilobium angustifolium:* **Chamerion danielsii,** !ISC
505. **Rorippa palustris** [subsp. **hispida**], !ISC, "Georgetown, July 20"
573. *Physalis pennsylvanica:* **P. virginiana,** !ISC, "Dome Rock, Aug. 8"
584. **P. heterophylla,** !ISC, "Denver, Aug. 13"
574. **Circaea alpina,** !ISC, "Dome Rock"
808. **Lupinus argenteus,** !ISC "Empire and Twin Lakes"

1881

Convolvulus sepium: **Calystegia sepium** subsp. **angulata,** !ISC, "S. Col"

1882

Aplopappus rubiginosus: **Machaeranthera phyllocephala,** !ISC, "Colorado" [using label headed "North American Pacific Coast Flora"]

Grindelia squarrosa, !ISC [using label headed "North American Pacific Coast Flora"]

Sporobolus ramulosus: **Muhlenbergia minutissima,** !ISC

1883

Bigelovia vaseyi: **Chrysothamnus depressus,** !ISC, "Gunnison Valley" [This is very common on high sagebrush areas on Black Mesa, above Sapinero.]

Gymnolomia multiflora: **Heliomeris multiflora,** !ISC, "Gunnison Valley"

Lithospermum: **Oreocarya suffruticosa,** !ISC, "Gunnison Valley"

Pinus edulis, !ISC, "Manitou" [El Paso County]

1888

Chamaerhodos erecta: **Aphanes occidentalis** [This specimen was collected in California! The label is of North America Pacific Coast Flora. The abbreviation COLO should be CAL]

Acnida: **Amaranthus arenicola,** !ISC, "Canyon of Arkansas" [three sheets]

Eriogonum lonchophyllum, !ISC, "Gunnison River Cañon, Sept. 24" [probably the Black Canyon]

E. watsonii: **E. gordonii,** !ISC, "Gunnison Valley" [probably near Grand Junction]

1889

Among the specimens are a number of collections made by General John Bidwell, a friend of Asa Gray. Some are from Colorado, but others, marked "Colo" were of taxa known only from California. Evidently the abbreviation was a mistake for "Cal." Bidwell is not listed in Ewan and Ewan (1981), but he seems to have been in the Colorado Springs area in 1889, possibly visiting General William Jackson Palmer, a pioneer railroadman who founded Colorado Springs in 1871 (Fisher 1939). Bidwell also evidently met Sir Joseph Hooker and Asa Gray on their trip to California in 1877. Fairchild (1939, p. 301) wrote: "Mr. Bidwell, one of the early pioneers in California, lived nearby [Chico] and had purchased a large tract of virgin lands containing many magnificent trees, including an oak supposed to be the finest in California. It was called the Hooker Oak in honor of the famous botanist, Sir Joseph Hooker, who had once visited Chico and expressed admiration for the tree."

Actaea spicata: **A. rubra** subsp. **arguta,** !ISC

Eriogonum inflatum, !ISC, "W Colo"

Lepidium: **L. alyssoides,** !ISC [probably collected near Colorado Springs]

Unnamed: **Atriplex corrugata,** !ISC, "W Colo" [probably collected near Grand Junction. Did he collect with Alice Eastwood here?]

A. wolfii, !ISC, "Gunnison, W Colo" [probably collected along Gunnison River near Grand Junction]

Chrysopsis: **Heterotheca villosa,** !ISC

Asclepias pumila, !ISC, "W Colo" [This is not reported for the Western Slope, where it is replaced by *A. subverticillata.*]

Eritrichium glomeratum: **Oreocarya thyrsiflora,** !ISC [mounted with Bidwell specimen and probably collected at the same place, vicinity of Colorado Springs]

Cryptogamic Collections (Lichens)[1] 9

It is not generally appreciated that Parry and Hall collected cryptogamic plants and that their collections were very important, especially to American lichenology. Tuckerman (1866, p. 4) wrote: "Other naturalists who have contributed to the extension of our list [of western American lichens] are . . . and Mr. E. Hall. To the careful observations of the latter, and to those, a season more recent, of Dr. Parry, almost the whole of our scanty knowledge of the important Alpine flora are confined." The specimens cited here are to be found in the Tuckerman collection of the Farlow Herbarium at Harvard University. I have not examined these personally. Hall also collected lichens in Illinois and Oregon (cf. *Lobaria hallii, Rinodina hallii*).

Biatora castanea: Tuckerman (1866, p. 24) **Bryonora castanea**

Biatora russula: Tuckerman (1888, p. 20) **Protoblastenia russula** [evidently the only Colorado record; not verified]

Biatora sanguineoatra: Tuckerman (1866, p. 24)

Buellia geographica: Tuckerman (1866, p. 26) **Rhizocarpon geographicum**

Buellia papillata: Tuckerman (1886, p. 26)

Buellia parasema: Tuckerman (1886, p. 26) [misidentification]

Cetraria cucullata: Tuckerman (1866, p. 12) **Allocetraria cucullata**

Cetraria islandica: Tuckerman (1866, p. 12)

Cetraria juniperina var. *terrestris:* Tuckerman (1866, p. 12) **Vulpicida tilesii**

Cetraria nivalis: Tuckerman (1866, p. 12) **Allocetraria nivalis**

Cetraria pinastri: Tuckerman (1882, p. 38) **Vulpicida pinastri**

Cladonia deformis: Tuckerman (1866, p. 23)

Cladonia furcata: Tuckerman (1882, p. 246)

Cladonia gracilis: Tuckerman (1866, p. 23): **Cladonia gracilis** or **C. ecmocyna**

1. Reported by Tuckerman (1864, 1866, 1882, 1888)

Dactylina madreporiformis: Tuckerman (1866, p. 12)

Evernia divaricata: Tuckerman (1866, p. 12)

Lecanora cinerea: Tuckerman (1866, p. 21): **Aspicilia sp.**

Lecanora pallescens: Tuckerman (1866, p. 20) [not mentioned by Tuckerman (1882, p. 196); probably *Ochrolechia upsaliensis*]

Lecanora schleicheri: Tuckerman (1866, p. 21) **Acarospora schleicheri**

Lecidea atrobrunnea: Tuckerman (1886, p. 24)

Lecidea fuscoatra: Tuckerman (1886, p. 24) [misidentification]

Lecidea morio: Tuckerman (1886, p. 25) **Sporastatia testudinea**

Lecidea vitellinaria: Tuckerman (1886, p. 24) **Carbonea vitellinaria**

Pannaria hypnorum: Tuckerman (1866 p. 17) **Psoroma hypnorum**

Parmelia conspersa: Tuckerman (1882, p. 64) **Xanthoparmelia sp.**

Parmelia molliuscula: Tuckerman (1882, p. 65) [Probably the specimen referred to is *Xanthoparmelia wyomingica.*]

Parmelia stygia: Tuckerman (1866, p. 13) [misidentification, possibly of *Pseudephebe minuscula;* not mentioned by Tuckerman (1882, p. 63)]

Parmelia tiliacea: Tuckerman (1866, p. 13) [misidentification]

Peltigera aphthosa: Tuckerman (1866, p. 16)

Peltigera venosa: Tuckerman (1866, p. 16)

Physcia stellaris: Tuckerman (1866, p. 13)

Placodium cerinum: Tuckerman (1866, p. 19) **Caloplaca cerina**

Placodium cladodes: Tuckerman (1864, 1866, p. 18) **Caloplaca cladodes**

Placodium elegans: Tuckerman (1866, p. 18) **Xanthoria elegans**

Placodium sinapispermum: Tuckerman (1866, p. 19) **Caloplaca sinapisperma**

Placodium vitellinum: Tuckerman (1866, p. 19) **Candelariella vitellina**

Solorina crocea: Tuckerman (1866, p. 16)

Stereocaulon tomentosum: Tuckerman (1866, p. 22)

Teloschistes chrysophthalmus: Tuckerman (1866, p. 13) [not found again in Colorado]

Umbilicaria cylindrica: Tuckerman (1866, p. 13) [misidentification of *U. virginis*]

Umbilicaria erosa: Tuckerman (1882, p. 86) [The description indicates that this was *U. torrefacta.*]

Umbilicaria hirsuta: Tuckerman (1866, p. 13)

Umbilicaria murina?: Tuckerman (1866, p. 13) [Tuckerman (1882, p. 87) listed this under *U. hirsuta* var. *grisea.*]

Umbilicaria proboscidea: Tuckerman (1866, p. 13) [not mentioned by Tuckerman (1882, p. 84); misidentification]

Umbilicaria vellea: Tuckerman (1882, p. 87) [The Colorado plant is now called **U. americana.**]

Urceolaria scruposa: Tuckerman (1866, p. 22) **Diploschistes scruposus**

Appendix A

New Taxa Described From Parry's Colorado Collections

Andropogon hallii Hackel, Sitzungsber. Akad. Wiss. Math. Naturw. (Wien) 89:127.1884

Antennaria margaritacea var. *subalpina* Gray, Proc. Acad. Nat. Sci. Philadelphia 15:67.1863

Arnica parryi A. Gray, Amer. Nat. 8:213.1874

Artemisia parryi Gray, Proc. Amer. Acad. Arts 7:361.1867 [1868]

A. scopulorum Gray, Proc. Acad. Nat. Sci. Philadelphia 15:66.1863

Asclepias hallii A. Gray, Proc. Amer. Acad. Arts 12:69.1877

Aster adscendens var. *parryi* D. C. Eaton, Bot. King's Exped. 139.1874 = *A. foliaceus* var. *parryi* Gray, Syn. Fl. 1(2):193.1884

A. pattersonii var. *hallii* Gray, Proc. Amer. Acad. 13:372.1878

Astragalus glabriusculus var. *major* Gray. Proc. Acad. Nat. Sci. Philadelphia 15:60.1863

A. hallii Gray, Proc. Amer. Acad. Arts 6:224.1864

A. lotiflorus var. *pedunculosus* and var. *brachypus* Gray, Proc. Amer. Acad. Arts 6:209.1864

A. microlobus Gray, Proc. Amer. Acad. Arts 6:203.1864

A. parryi Gray, Amer. J. Sci. (2)33:410.1862

A. sparsiflorus Gray, Proc. Amer. Acad. Arts 6:205.1864

A. sparsiflorus var. *majusculus* Gray, loc. cit.

A. tridactylicus Gray, Proc. Amer. Acad. Arts 6:527.1865

Bigelowia howardii Parry ex Gray, Proc. Amer. Acad. Arts 8:641.1873

Boltonia latisquama Gray, Amer. J. Sci. 33:328.1862

B. grandiflora var. *minor* Gray, Proc. Acad. Nat. Sci. Philadelphia 15:67.1863

Campanula parryi Gray, Synopt. Fl. 2(1):395.1866

Carex bonplandii var. *minor* F. Boott, Proc. Acad. Nat. Sci. Phila. 1863:77.1863

C. hallii Olney, Hayden Surv., 1871, p. 496

C. oreocharis Holm, Amer. J. Sci. IV.9:358.1900

C. violacea C. B. Clarke, Kew Bull., Add. Ser. 8:87.1908 [erroneously attributed to California]

Cirsium acaule var. *americanum* Gray, Proc. Acad. Nat. Sci. Philadelphia 15:68.1863

C. eriocephalum Gray, *non* Wallr., loc. cit., p. 69

Claytonia arctica var. *megarhiza* Parry *ex* Gray, Amer. J. Sci. (2)33:406.1862

Cnicus carlinoides var. *americanus* Gray, Proc. Amer. Acad. Arts 10:48.1874

C. parryi Gray, loc. cit., p. 39

Cymopterus alpinus Gray, Amer. J. Sci. (2)33:408.1862

C. anisatus Gray, Proc. Acad. Nat. Sci. Philadelphia 15:63.1863

Danthonia parryi Scribn., Bot. Gaz. 21:133.1896

Draba streptocarpa Gray, Amer. J. Sci. (2)33.242.1862

Echinocactus simpsonii var. *minor* Engelm. Trans. Acad. Sci. St. Louis 2:197.1863

Erigeron grandiflorum var. *elatius* Gray, Amer. J. Sci. (2)33:237.1862

Gentiana affinis var. *brachycalyx* Engelm. *ex* Gray, Proc. Acad. Nat. Sci. Philadelphia 15:74.1863

G. barbellata Engelm., Trans. Acad. Sci. St. Louis 2:216.1866

G. parryi Engelm., Trans. Acad. Sci. St. Louis 2:218.1863

Geranium fremontii var. *parryi* Gray, Amer. J. Sci. 33:405.1862

Gilia globularis Brand, Pflanzenr. IV, 250, Heft 27:120.1907

Graphephorum? flexuosum Thurb. *ex* Gray, Proc. Acad. Nat. Sci. Philadelphia 15:78.1863

Haplopappus croceus Gray, loc. cit., p. 65, 1863

H. parryi Gray, Amer. J. Sci. (2)33:239.1862

Helenium hoopesii Gray, loc. cit.

Helianthella parryi Gray, loc. cit.

Heuchera hallii Gray, loc. cit., p. 62

Juncus hallii Engelm, Trans. Acad. Sci. St. Louis 2:446.1866

J. parryi Engelm., loc. cit.

Lappula subdecumbens Parry, Proc. Davenp. Acad. 1:148.1876

Linosyris parryi Gray, Proc. Acad. Nat. Sci. Philadelphia 15:66.1863

Lupinus ornatus var. *glabratus* S. Wats. in Porter, Synopsis Fl. Colo. 19.1874

Lychnis montana S. Wats., Proc. Amer. Acad. Arts 12:247.1877

Lygodesmia juncea var.? *rostrata* Gray, Proc. Acad. Nat. Sci. Philadelphia 15:69.1863

Melica hallii Vasey, Bot. Gaz. 2:296–297.1881

Mimulus hallii Greene, Bull. Calif. Acad. Sci. 1:113.1885

Muhlenbergia pungens Thurb. *ex* Gray. Proc. Acad. Nat. Sci. Philadelphia 15:78.1863

Neoparrya lithophila Mathias, Ann. Mo. Bot. Gard. 16:393.1929

Nuphar polysepalum Engelm., Trans. Acad. Sci. St. Louis 2:282–285.1867

Oxytropis parryi Gray, Proc. Amer. Acad. Arts Sci. 20:4.1884

Paronychia pulvinata Gray, Proc. Acad. Nat. Sci. Philadelphia 15:53.1863
Pedicularis parryi Gray, Amer. J. Sci. (2)34:251.1862
P. procera Gray, loc. cit.
P. scopulorum Gray, Synopt. Fl. 2(1):308.1886
Penstemon fremontii var. *parryi* Gray *ex* S. Wats., Bot. King's Exped. 218.1871
P. glaucus var. *stenosepalus* Gray, Proc. Amer. Acad. Arts 6:760.1862
P. hallii Gray, Proc. Amer. Acad. Arts 6:70.1862
P. harbourii Gray, loc. cit., p. 71
Phlox caespitosa var. *condensata* Gray, Proc. Amer. Acad. Arts 8:254.1879
Picea pungens Engelm., Gard. Chron. (2)11:334.1879
P. engelmannii Parry *ex* Engelm., Trans. Acad. Sci. St. Louis 2:212–214.1863
Pinus aristata Engelm., loc. cit., pp. 205–207
Placodium cladodes Tuck., Proc. Amer. Acad. Arts Sci. 6:382–383.1864
Plagiochasma erythrospermum Sull. *ex* Aust., Proc. Acad. Nat. Sci. Phila. 21:229.1869 [1870]
Polemonium confertum Gray, Proc. Acad. Nat. Sci. Philadelphia 15:73.1863
P. confertum var. *mellitum* Gray, Proc. Acad. Nat. Sci. Philadelphia 15:73.1863
P. foliosissimum Gray, Proc. Amer. Acad. Arts 8:281.1870
Primula parryi Gray, Amer. J. Sci. (2)34:257.1862
Ranunculus adoneus Gray, Proc. Acad. Nat. Sci. Philadelphia 15:56.1863
Sedum rhodanthum Gray, Amer. J. Sci. (2)33:405.1862
Senecio amplectens Gray, loc. cit., p. 240
S. amplectens var. *taraxacoides* Gray, Proc. Acad. Nat. Sci. Philadelphia 15:67.1863
S. aureus var. *alpinus* Gray, Amer. J. Sci. 33:249.1862
S. aureus var. *(alpinus) werneriaefolius* Gray, Proc. Acad. Nat. Sci. Philadelphia 15:68.1863
S. bigelowii var. *hallii* Gray, Proc. Acad. Nat. Sci. Philadelphia 15:67.1863
S. cernuus Gray, *non* L., Amer. J. Sci. 33:239.1862
S. crocatus Rydb., Bull. Torr. Bot. Club 24:299.1897
S. soldanella Gray, Proc. Acad. Nat. Sci. Philadelphia 15:67.1863
Silene hallii S. Wats., Proc. Amer. Acad. Arts 21:446.1886
Stipa porteri Rydb., Bull. Torr. Bot. Club 32:599.1905
Synthyris alpina Gray, Amer. J. Sci. (2)34:251.1862
Talinum pygmaeum Gray, Amer. J. Sci. (2)33:407.1862
Thalictrum scopulorum Greene, Lfl. Bot. Obs. Crit. 2:91–92.1910
Thaspium trachypleurum Gray, Proc. Acad. Nat. Sci. Philadelphia 15:63.1863
Townsendia sericea var. *leptotes* Gray, Proc. Amer. Acad. Arts 16:85.1880
Trifolium parryi Gray: Amer. J. Sci. (2)33:409.1862
Trollius laxus var. *albiflorus* Gray, Amer. J. Sci. 33:241–242.1862

Vaseya comata Thurber in Gray, Proc. Acad. Nat. Sci. Philadelphia 15:79–80.1863

Vesicaria montana Gray, loc. cit., p. 53

Woodsia scopulina D. C. Eaton, Canad. Naturalist & Quart. J. Sci. (2)2:90.1865

Appendix B

The Parry Herbarium

In July 1891, Mrs. E. R. Parry (C. C.'s wife) offered the Parry collection and scientific library for sale (Mrs. E. R. Parry 1891). It contained 18,000 specimens, of which 15,800 were North American and 1,430 Mexican. Other important details of the ultimate disposition of the Parry herbarium are given by Gabel (1981). Parry's herbarium now resides in the Ada Hayden Herbarium of Iowa State University, Ames. It is kept separate as a historical archive, which makes its contents very easily accessible.

Evidently Mrs. Parry retained control over the collection even after C. C. Parry deposited it in the Davenport Academy. C. C. Parry's account of his donation follows.

October 28th, 1878. — Regular Meeting

Dr. R. J. Farquharson, President, in the chair.

Thirty-two members and visitors present.

The President reported the actions of the trustees during the month past, and presented the following letter from Dr. Parry, giving an account of the botanical collection which he has deposited in the botanical room of the Academy.

To the Trustees of Davenport Academy of Natural Sciences:

Gentlemen: Your courteous invitation to deposit my botanical collection in the commodious room of the Academy assigned to that department has been complied with, so far as the incomplete arrangement of the material, mainly the result of thirty years' active work, would allow.

It may be proper in this connection to state briefly the character of this collection, and the principal sources from which it has been derived.

My earliest gatherings in the botanical field were begun in 1842, while residing in the attractive floral district of northeastern New York, and continued more or less actively for five years, while occupied in a course of medical studies. During this interval I spent one season in central New York, including a trip to Niagara Falls. The two last years of this period was especially memorable by being favored with the personal acquaintance of the distinguished American botanist, Dr. John Torrey, to whose assistance and encouragement, equally shared by nearly all active American botanists of this generation, I am largely indebted for whatever success I may have attained.

In the fall of 1846, I removed to Davenport, Iowa, and the season following, I was actively engaged in securing the flora of this district, including a summer excursion to central Iowa, in the vicinity of the present State Capital, Des Moines, with a United States land surveying party under the charge of Lieut. J. Morehead.

In 1848 I was connected with Dr. D. D. Owen's geological survey of the North-West, making botanical collections along the course of the St. Peters River and up the St. Croix as far as Lake Superior. A list of the plants collected during this and the preceding season was included in Dr. Owen's report published in 1852.

In 1849 I was appointed botanist to the Mexican Boundary Survey, going by way of the Isthmus of Panama to San Diego, California, which latter place was reached in July. In September of the same year I accompanied an astronomical party to the junction of the Gila and Colorado rivers, returning to San Diego in December. The important collections of this season were unfortunately lost in crossing the Isthmus of Panama while in charge of the late Gen. A. W. Whipple, being probably involved in a disastrous fire while stored in Panama awaiting transportation.

In the subsequent year, 1850, this loss was partially made up by somewhat extensive collections in the vicinity of the Southern Boundary line, and including a land trip up the coast as far as Monterey.

In the year 1851 I was ordered to Washington to make up my report, but before concluding it I was unexpectedly summoned to join the field party on the survey of the boundary then transferred to El Paso on the Rio Grande. This point was reached by an overland trip *via* San Antonio, Texas, late in the fall of that year (1851). In January of the follwing year (1852) I was connected with a small detailed party of exploration across the country west of El Paso, extending as far as the Pima settlements on the Gila River, returning by the same route to El Paso in April. Subsequently I was connected with various surveying parties on the line of the Rio Grande south of El Paso, including late in the season the section of the river below Presidio del Norte, comprising a succession of gigantic chasms which never before or since have been visited by any botanist.

In the winter of 1852–1853 I returned to Washington and made up my report, since published in the bulky volumes of the Mexican Boundary Survey. The interval from 1854 to 1860 was spent mainly in Davenport, not actively engaged in botanical work.

In the spring of 1861 the culmination of the Pike's Peak fever again opened the way for western exploration, and in a private collecting trip to the Rocky Mountains, I succeeded in securing a rare collection of alpine plants including, among many novelties, some of the early discoveries of Dr. James on Long's expedition in 1820. In the following season I was associated with E. Hall and J. P. Harbour in further exploration of the Rocky Mountain district, the botanical results of which were published in Proceedings of the Philadelphia Academy for 1863.

In 1864, in company with Dr. J. W. Velie, then of Rock Island, Ill., I continued my Rocky Mountain coillections, embracing the districts of Long's Peak and Middle Park.

In 1867, I accompanied a railroad surveying party in the interests of the Pacific Railway Company across the continent on the line of 35° parallel north latitude. The most valuable part of my collections during that season were made in western Kansas and southeastern Colorado, passing by the Sangre de Cristo Pass to northern New Mexico; thence late in the winter season through Arizona, crossing the Sierra Nevada at Tehachapi Pass, and through the Tulare and San Joachin valleys to San Francisco. A list of the plants comprised in this collection was subsequently published in Dr. W. A. Bell's work entitled *New Tracks in North America,* but without an opportunity for personal revision by the collector.

An interval of several years subsequent to the latter trip was occupied in filling the position of Botanist to the Agricultural Department in Washington. The principal work there devolving upon me was that of arranging the extensive botanical collections which, as the result of various government explorations, had accumulated at the Smithsonian Institution. The bulk of these had previously passed through the hands of Dr. Torrey, whose gratuitous labors in reducing this mass of raw botanical material to systematic shape have never yet been properly acknowledged [see Appendix D].

On being relieved from this position in the fall of 1871, the season following I again revisited the Rocky Mountain alpine district, being then accompanied for the first time by our associate, J. Duncan Putnam.

In 1873 I was attached to the North-Western Wyoming Expedition, under Capt. W. A. Jones, extending through the Wind River District to the Yellowstone National Park, Mr. Putnam being assigned as my meteorological assistant.

In 1874 I made a private collecting tour to south Utah, securing a valuable collection of the flora of the singular desert district in the valley of the Virgen, near St. George. [Parry was joined by Edward Palmer on this trip and the following one of 1875. See C. C. Parry (1876).]

In 1875, again accompanied by Mr. Putnam [and Edward Palmer], I spent the summer in central Utah in the vicinity of Mount Nebo. In the fall of that year I continued my collecting trip to southern California, and in the season of 1876, in connection with Prof. J. G. Lemmon, the enthusiastic California botanist, made a very full collection of the plants in the vicinity of San Bernardino, including the high mountain district adjoining, and the desert stretches lying east of the Sierra Nevada.

My last and closing labors as botanical collector [with Edward Palmer] were made during the present season, mainly in the vicinity of San Luis Potosí, Mexico, extending on my return trip by way of Saltillo and Monterey to the more familiar botanical district of western Texas, which I had partly explored twenty-six years previous.

From all these various sources, collections, more or less complete, have accumulated on my hands, the great bulk being fortunately distributed far and wide to the different herbaria of America and Europe. An active correspondence with the principal American botanists during the past thirty years has added largely, in the way of exchanges, to the material for illustrating Western American Botany. Hoping only for an opportunity to reduce this scattered material to systematic order, and to see it safely deposited in some scientific institution in the West, where it properly belongs, I gladly avail myself of the invitation extended to me by the Trustees of the Davenport Academy of Natural Sciences.

In fully realizing the fact that with advancing years my active labors as a collector and explorer are virtually finished, it is a pleasant reflection that some of the results of my labors, here deposited in an Academy of Science with which I have been from the first identified, and located in my adopted home on the west bank of the Mississippi, may perchance prove a source of assistance and encouragement to future botanists long after the "gathering hand" shall itself be "gathered."

Appendix C

Parry's Dismissal as Botanist for the Department of Agriculture

Parry was summarily dismissed from the Department of Agriculture in 1871. In an obscure paper largely forgotten and never cited by biographers, Asa Gray (1872) published some pertinent correspondence in protest. The letters shed some light on the characters of the protagonists and especially on that of Parry himself that I feel are worth including in this memoir. Gray's gift for subtle sarcasm is noteworthy.

Correspondence Relating to the Dismissal of the Late Botanist to the Department of Agriculture at Washington

Editors, American Naturalist. Dear Sirs:—I have to request that you will place before your readers of the *AMERICAN NATURALIST* the correspondence here enclosed.

Dr. Parry was thought to have performed the duties of Botanist to the Department of Agriculture to the entire satisfaction of the previous Commissioner. His extraordinarily abrupt dismissal upon the incoming of the present Commissioner, following a course of vexatious treatment to which, he states, he was subjected by his Chief Clerk, does not seem calculated to win the confidence of scientific men in the present administration of a department in which they naturally feel much interest.

Very respectfully yours,
Asa Gray

Department of Agriculture,
Washington, D.C., September 27, 1871

Hon. F. Watts, *Commissioner of Agriculture*

Sir:—In order to enable me to comply strictly with the regulations of this Department in regard to ordinary correspondence in connexion with my official duties as botanist, I respectfully ask to be furnished with written instructions on the following points. 1st. Should letters addressed to me personally, as botanist of the department, imparting or requesting information on botanical subjects, be answered and signed by me personally as botanist, or in the name of the Commissioner? 2nd. In sending botanical specimens to be named, or in returning such as have been sent to me to name, should the accompanying letter be signed by myself as botanist or by the Commissioner?

Having heretofore exercised my own discretion in this matter, with due regard to the scientific interests of the department and to facilitate the business of my division, I desire to avoid any future misunderstanding by receiving definite written instructions on these points for my guidance.

Respectfully yours,
C. C. Parry, *Botanist, Agr. Dept.*

Department of Agriculture
Washington, D.C., September 27, 1871

C. C. Parry, *Esq., Washington, D.C.*

Sir:—Your services as botanist of this Department will not be required after this date.

I am respectfully,

Frederick Watts, *Commissioner*

Department of Agriculture
Washington, D. C., September 27, 1871

Hon. Frederick Watts, *Commissioner of Agriculture*

Sir:—I have the honor to acknowledge your letter of this date informing me that my "services as botanist of this Department will not be required after this date," for which I sincerely thank you.

I respectfully request that you will designate some person from the department tomorrow to be with me in selecting my private property, books, etc. from that belonging to the Department,

Respectfully yours,
C. C. Parry, *Botanist, Agr. Dept.*

To the Honorable Judge Watts, *U. S. Commissioner of Agriculture, Washington*

The undersigned, botanists, well acquainted with Dr. C. C. Parry, and having a high opinion of his ability, industry, entire probity and honorable character, as well as of his peculiar qualifications for the position, acting upon their view of the best interests of the science they represent, and sincerely believing that his dismissal must have taken place under some misapprehension, hereby respectfully solicit that the Commissioner would take into consideration the propriety of reappointing Dr. Parry to the position of Botanist in the Department of Agriculture.

John Torrey,
Asa Gray,
Wm. H. Brewer, *Prof. Agriculture in Yale College*
Daniel C. Eaton, *Prof. Botany in Yale College*

Harvard University Herbarium, November 22, 1871

P.S.—A copy is being forwarded to Messrs. Watson, Engelmann, and Canby, for their signatures.

Department of Agriculture
Washington, D.C., November 27, 1871

To Prof. Asa Gray.

Dear Sir:—Prof. Henry this morning placed in my hands the note of Profs. Torrey, Brewer, Eaton, and yourself, asking me "to consider the propriety of reappointing Dr. Parry to the position of Botanist in the Department of Agriculture." The respect which I must necessarily have for a suggestion coming from such a source induces me carefully to review my action in the matter of Dr. Parry's removal; and my conclusion is, that my own self respect and especially the interests of this Department, forbid that I should reverse that which I did in care and reflection. I did not, to Dr. Parry himself, assign any reason for his removal, simply because he did not afford me any opportunity to do so. I did not see him afterwards, but I should add that it was quite acceptable to me that I was not called upon to assign reasons which it would have been as disagreeable to me to utter as for him to hear. Nor do I now desire to say any thing about Dr. Parry that might disparage him in the estimation of his friends.

I am, most respectfully, your ob't. servant
Frederick Watts

Cambridge, Mass., November 30th, 1871

To the Hon. Frederick Watts, *U. S. Commissioner of Agriculture*

Sir:—I have to acknowledge your favor of the 27th inst. in reply to the memorial addressed to you by Professors Torrey, Brewer, Eaton, and myself. It still appears to me that the friends of Dr. Parry are entitled to know the reason of his summary dismissal by you,—all the more so that your letter intimates, without directly asserting, some moral delinquency on his part. I am still so confident that you must have been misled, that I respectfully ask leave to print your letter to me along with the memorial to which it is a reply, in case you still decline to furnish the charges upon which Dr. Parry's dismissal was grounded.

I am, sir, respectfully, your obedient servant,

Asa Gray

Department of Agriculture,
Washington, D. C., Nov. 8, 1871

To Professor Asa Gray:—Yours of the 30th of November was handed to me yesterday by Professor Henry. If it were not that you say that my former letter to you "intimates without directly asserting some moral delinquency" on Dr. Parry's part, I would content myself by saying that my judgment dictated to me the propriety of Dr. Parry's removal. But I have concluded to put you in possession of the whole subject.

When I took charge of this Department, my first duty was to look into and to understand the divisions of subjects which make up its whole, the work that had been done, and the character and competency of each individual who had charge of that work. Among the divisions was that of the Botanist, with Dr. Parry in charge of it. My attention was called to the inquiry, how and to what extent the work of this division conduced to the practical operations of the Department. I found that nothing at all had been done by Dr. Parry beyond his attention to the preservation of the herbarium. This Department is designed to render the development and deductions of science directly available to practice, that farmers and horticulturists may be benefited by them. The principles of vegetable physiology, their relations to climate, soils, and the food of plants, and the diseases of plants, which are principally of fungoid origin, it is clearly the duty of a botanist to investigate. If possible, he should throw some light upon the origin and condition of growth of the lower orders of cryptogamic botany. This is a domain into which I could not discover that Dr. Parry had ever entered, so far as his practical work here gave any indication. The routine operations of a mere herbarium botanist are practically unimportant.

In the course of my investigation, my attention was also drawn to letters written by Dr. Parry, which I deemed objectionable because of his mode of expression, wanting in perspicuity and not creditable to the Department. These things, and what I also learned that my predecessor had signified to Dr. Parry, to the effect that his letters should be submitted to him and for his signature before they were sent away, induced me to direct my chief clerk to have a conversation with Dr. Parry, and to say to him that, as the head of the Department, I was responsible for all that emanated from it, and that all letters on official business must be sent open to me, for my signature and frank. I returned with this message a sealed package for which my frank was asked. At another time I returned to Dr. Parry by my chief clerk, a letter which he had written and which I did not think proper should be sent, and which the Doctor passionately tore up and threw into the waste basket. This he subsequently apologized for to the gentleman he had thus insulted. On the 25th of September, after these various conversations between my chief clerk and Dr. Parry, he wrote another letter addressed to "My Dear Doctor." It had no other designation. For whom it was intended, I did not learn, or if I did I have forgotten. It concluded, "yours, 'officially,' C. C. Parry." I wrote on this last letter, "This is not very intelligible in its last sentence; besides, the Botanist can sign no official letters. What his "official" means I do not understand, but under the circumstances, I think it is intended for impertinence. It then occurred to me that I would dismiss Dr. Parry, but held the matter under advisement for two days, until the 27th of September, when I received a note from him, in which he requested me to furnish him with *written* instructions (underscoring the word), and which contained two queries respecting letters from the Department. I did not think he was in want of the information he asked for, and my answer to his note was that the Department did not longer require his services. My conviction was then, and is now, that whatever may be the qualifications of Dr. Parry as a botanist, he was not competent creditably to discharge the duties which should devolve upon him in connection with this Department, and therefore, without passion or prejudice, I determined to dismiss him.

A word in reply to your suggestion about printing my letter and your memorial. I decline to be a party myself to any such proceedings. But if you will take the whole responsibility of it, I shall never complain that you have violated a confidence which I never intended to impose.

I am, very respectfully,

Frederick Watts, *Commissioner of Agriculture*

Botanic Garden
Cambridge, Mass., December 11th, 1871

To the Hon. Frederick Watts, *U. S. Commissioner of Agriculture, Washington, D.C.*

My Dear Sir:—I have to thank you for your letter "Nov." [Dec.] 8th, in response to mine of Nov. 30.

You will permit me to remark, that the dismissal, without an hour's notice, of Dr. Parry from a position the duties of which he was thought to have performed acceptably to your predecessor, must of itself, if unexplained, cast an injurious reflection upon his character or conduct. Then your letter in reply to the memorial which solicited his recall, stating that the reasons for such dismissal were of a nature which it would have been as disagreeable for you to utter as for him to hear, and that you "do not now desire to say anything about Dr. Parry which might disparage him in the estimation of his friends,"—all this certainly conveyed to my mind the conviction that some serious delinquency had been charged. It is with satisfaction, therefore, that I have read your letter now before me, obligingly written "to put [me] in possession of the whole subject." I learn from it that the reasons for Dr. Parry's summary and ignominious dismissal relate to some details of form in the mode of conducting official botanical correspondence,—to a momentary loss of temper in the presence of one of your subordinates (evinced by the mode in which he destroyed a letter of his which had been returned to him to be cancelled), and for which he duly apologized,—to the subscribing of a letter addressed familiarly "My Dear Doctor" [evidently some botanical correspondent] by the phrase "yours officially,"—that in some letters you found "his mode of expression wanting in perspicuity" (a fault into which more practised writers may sometimes fall),—and finally, that you did not discover in Dr. Parry the kind or degree of botanical qualifications for the post which you were entitled to expect, and deemed the services of "an herbarium botanist" practically unimportant.

As your letter had relieved my own mind from a painful anxiety upon this subject, it may have the same effect upon others, upon whose minds also your action had left the alternative of supposing either bad conduct on the part of one hitherto highly esteemed, or of very hard usage towards him (it was thought through some misrepresentation of him or some misapprehension of yours). I think it proper and just, therefore, to make use of the permission you grant, and to take the responsibility of making public, in scientific circles, first, the correspondence between Dr. Parry and yourself, and second, that between ourselves.

I am, very respectfully yours,

Asa Gray

Appendix D
Publications of C. C. Parry

*(except those listed in References Cited)**

Parry published relatively few papers in scientific journals but a great number of articles in newspapers. The latter venue is generally considered to be beyond the pale for scientists, but in Parry's day, this was a legitimate means of communicating with the scientific and lay public. In these articles the spirit of the scientist emerges in a way that is impossible in formal publication. Natural history magazines such as *Smithsonian, Nature,* and *Natural History* are slightly more acceptable but were unknown on the U.S. frontier. Most of the old naturalists were confirmed amateurs and did not consider the lay public to be below them. The titles of Parry's articles are indicative of his very wide interest in social as well as scientific matters. By stepping in fancy into Parry's shoes, one may still read Parry's discourses with some profit.

American Association for the Advancement of Science, Proceedings

The habitative features of the North American Continental Plateau near the line of the thirty-ninth parallel of North Latitude. 17:312

The Rocky Mountain alpine region. 17:248

On a form of boomerang in use among the Moqui Pueblo Indians of North America. 20:397

American Naturalist

Visit to the original locality of the new species of *Arceuthobium* in Warren County, N.Y. 6:404

Botanical observations in western Wyoming, with notices of rare plants and descriptions of new species collected on the route of the

* Adapted from Mrs. E. R. Parry (1893)

Northwestern Wyoming Expedition under Capt. W. A. Jones. 8:9, 102, 175, 211

Herbarium cases. 8. 1874

Botanical observations in southern Utah. 9:114, 139, 199, 267, 346

Botanical Gazette

On some recent notes and descriptions of *Eriogoneae* in *Proceedings of California Academy of Science.* 2:54–56

British Association for the Advancement of Science. Report.

North American desert flora between 32°–42° North Latitude. 40:122 (published also in *J. Bot.* 8:343–346. 1870)

California Academy of Science, Bulletin

Pacific Coast alders. 2:351–354

California manzanitas. A partial revision of the *Uva Ursa* section of the genus *Arctostaphylos* Adanson, as represented on the North American coast. 2:483–496

Chicago Evening Journal

The Far West: Its external features and natural resources. Dec. 8, 1863 (also *Davenport Gazette,* Dec. 17, 1863)

Eastern Nebraska along the Platte Valley. Dec. 15, 1863

Western Nebraska and eastern Colorado to the base of the Rocky Mountains. Dec. 22, 1863 (also *Davenport Gazette,* Dec. 31, 1863)

External features and natural resources: First impressions of Rocky Mountain scenery. Dec. 30, 1863 (also *Davenport Gazette,* Jan. 7, 1864)

The Snowy Range of the Rocky Mountains. Jan. 6, 1864 (also *Davenport Gazette,* Jan. 9, 1864.)

The passes of the Snowy Range of the Rocky Mountains leading to Middle Park. Jan. 15, 1864 (also *Davenport Gazette,* Jan. 18, 1864)

The route of the Pacific Railroad: Its external features and natural resources. Jan. 20, 1864

The Far West: Its present wants and future prospects. Jan. 27, 1864 (also *Davenport Gazette,* Jan. 29, 1864)

A city in the heart of the Rocky Mountains: Its history, scenery, altitude, surroundings, etc. Dec. 21, 1864

Dividing line between the agricultural and pastoral regions (date uncertain)

Denver City (date uncertain)

Midsummer week in Middle Park. Sept.(?) 1864

Far West sketches: Ascent of Long's Peak. Sept. 18, 1864 (also *Davenport Gazette,* Oct. 2 (?), 1864.)

Colorado Chieftain

Letter from Pueblo, Colorado. Oct. 21, 1869

Commission of Inquiry to the Island of Santo Domingo, 1871. Forty-second Congress, First Session, Senate Executive Document No. 9. Report

Report on the botanical features, agricultural products, and timber growth of the Peninsula of Samana, Santo Domingo. p. 71

Botany of the Southern District of Santo Domingo. p. 86

Davenport Academy of Sciences, Proceedings

Valedictory address. 1:19

Obituary notice of Prof. John Torrey. loc. cit., p. 44

Annual address. loc. cit., p. 67

Summer botanizing in the Wasatch Mountains, Utah (letter to Prof. Asa Gray). loc. cit., p. 145.

A new California lily. loc. cit., p. 188, pl. 5, 6c

Annual address (part omitted from vol. 1). loc. cit., p. 355

Notice of the late I. A. Lapham. loc. cit., p. 29

Ode on laying the Corner-stone. loc. cit., p. 178

On depositing the Parry botanical collection. loc. cit., p. 279 (autobiographical)

Oxytheca. Two new species from Southern California. 3:174

Biographical sketch of the late J. Duncan Putnam. loc. cit., p. 255

Arctostaphylos, Adans. Notes on the United States Pacific Coast species, including a new species from Lower California. 4:31–37

New plants from Southern and Lower California. loc. cit., pp. 38–40

Chorizanthe, R. Brown: Revision of the genus, and rearrangement of the annual species—with one exception, all North American. loc. cit., pp. 45–65

Obituary notice of Dr. John LeConte. loc. cit., p. 230

Obituary notice of Dr. George Engelmann. loc. cit., p. 242

Harfordia, Greene & Parry. A new genus of Eriogoneae from Lower California. 5:26–28

Lastarriaea, Remy. Confirmation of the genus, with character extended. loc. cit., pp. 35–36

The North American genus *Ceanothus,* with an enumerative list, and notes and descriptions of several Pacific Coast species. 5:162–174

Chorizanthe, R. Brown. Review of certain species heretofore improperly characterized or wrongly referred, with two new species. loc. cit., pp. 174–176

Memorial of Prof. David S. Sheldon. loc. cit., p. 179

Ceanothus, L. Recent field notes, with a partial revision of species. loc. cit., p. 185–194

Early reminiscences of Richard Smetham. loc. cit., p. 205

Davenport Daily Democrat

English rural life. Mar. 22, 1885

Davenport Gazette

Panama—New Grenada, S. A. July 5, 1849

Account of the Isthmus from Chagres to Panama. May 23, 1849

A public library. Apr. 1, 1854

Geology of Iowa. Nov. 15, 1858

Claims of Dr. Fountain. Jan. 31, 1862 (read before the Scott County Medical Society)

The utility of collections in natural history in connection with common schools. April (?) 1864 (high school lecture)

Peat as an article of fuel. Jan. 25, 1866

Peru coal in Davenport. Nov. 19, 1866

St. Louis: Its general aspect, street railroads, Nicholson pavements, etc. 1867

Departure on an exploring expedition to the Pacific. May 8, 1867

Indian war and the Pacific Railroad. July 19, 1867

Sangre de Cristo Pass in the Rocky Mountains. September 1867

From New Mexico: Harvest month on the Rio Grande. Nov. 6, 1867

From Arizona: An interesting account of the Territory. Jan. 6, 1868

The Great Colorado of the West: Its navigable waters and deep cañons. Feb. 11, 1868

Davenport Academy of Science. June 1, 1868
General Kit Carson. June 12 (?), 1868
New England seen through western eyes. Aug. 1868
Aspects of Rocky Mountain scenery. Apr. 27, 1868 (lecture)
Letter from Washington. May, 1869
Letter from California: Climatic luxuriance. Jan. 21, 1876
Letter from Philadelphia. June 13, 1877
Letter from Cambridge, Mass. July 14, 1877
Mexico: Aztec capitol, etc. Feb. 13, 1878
Mexico: A winter's journey and its observations. Mar. 1878
Mexico and Mexicans impartially presented. May 22, 1878
Thirty-six years a botanist. Collection deposited in the Davenport Academy of Natural Sciences. Oct. 26, 1878
The potato—was it given to the world by the Aztecs? Nov. 3, 1879
Shall we have a state entomologist? Mar. 12, 1880
Northern Oregon: A trip on the Columbia River. Aug. 27, 1880
The Columbia: The people and the valley. Sept. 9, 1880
The Gulf of California: A journey thither by an old friend. Dec. 4, 1880
Midwinter in Southern California. Feb. 7, 1881
Letter from Los Angeles. June 2, 1881
On the Pacific Coast: The indescribable Yosemite, etc. Aug. 23, 1881
The autumn in the Pacific States: Its peculiar aspects. Dec. 1, 1881
A parting paper. Last address to the D.A.N.S. before departure for Europe. Apr. 17, 1884

Geological Survey of Wisconsin, Iowa, and Minnesota, by David Dale Owen.

Systematic catalogue of plants of Wisconsin and Minnesota. Art. V, pp. 606. 1848

Iowa Instructor and School Journal

Claims of natural science as a branch of mental education. vol. 1, nos. 3–9. 1859
Evil effects resulting from an improper association of educational interests with party politics. vol. 2, no. 2
The three periods in a writer's progress. vol. 4, no. 9

Kansas Pacific Railway Survey, 1867–1868. Report.

Preliminary Report, and extract from detailed report, as geologist and naturalist, pp. 196 and 203

Madroño

From San Diego to the Bay of All Saints, Lower California, and back: Notes of a botanist visiting Mexican soil. 1:218–221. 1929 (reprinted from *San Francisco Bulletin,* Apr. 28, 1882)

Oscaloosa (Iowa) Herald

Report on botany and horticulture. Jan. 1877 (read before the State Horticultural Society)

Overland Monthly

Early botanical explorers of the Pacific coast. Oct. 1883
Rancho Chico. June 1888

Pamphlets (privately published)

Historical address on the early exploration and settlement of the Mississippi Valley (delivered at Davenport, Iowa, June 21, 1873)
What a botanist saw in Europe (read before the Chautauqua Literary and Scientific Circle at Monterey, Calif., July 1887)

Rocky Mountain News

Mountain heights in the United States. Mar. 19, 1863

St. Louis Academy of Sciences, Transactions.

Notice of some additional observations on the physiography of the Rocky Mountains, made during the summer of 1864. 2:272
Account of the passage through the Great Cañon of the Colorado by Mr. James White in 1867, with geological notes. loc. cit., no. 3

San Bernardino Times

Rocky Mountain cacti (probably 1876)
On botany of California (date uncertain)

San Diego Union

Address before Natural History Society. Mar. 12, 1882
Remarks on Lower California. Feb. 9, 1883 (address before Natural History Society)

San Francisco Bulletin

San Bernardino. Two struggling systems; Growth of civilization; The gardens of Riverside; The hopeless region of the Colorado. Feb. 16, 1881
The desert palm. Botanical history of the *Washingtonia filifera,* its character and habitat. Mar. 24, 1881
San Joaquin Valley. Past, present, and future of a vast region. Sept. 7, 1881
Tree culture. Dec. 23, 1881
The desert ironwood. Explorations in the Mojave and Arizona regions. Jan. 12, 1882
Coahuila Indians. Feb. 1, 1882
Fruit lands. Irrigation from the Santa Anna River. Mar. 9, 1882
San Diego revisited. Mar. 28, 1882
Lower California. Notes of a botanist visiting Mexican soil. Apr. 28, 1882
Down south. May 15, 1882
A botanist in Lower California. Mar. 15, 1883
Across the Mojave Desert. Apr. 12, 1883
Howard Schuyler. Jan. 23, 1884

Torrey Botanical Club, Bulletin

Fruit of *Cucurbita.* vol. 9, p. 30, pl. 14
A new North American rose. loc. cit., p. 97
A new species of *Oxytheca.* vol. 10, p. 23
Cucurbita californica. loc. cit., p. 50, with plates
Note on *Harfordia,* Greene & Parry. vol. 16, p. 277

United States and Mexican Boundary Survey. 1859. Report

Reconnaissance to the mouth of the Gila River from San Diego, Calif., Sept. 11 to Dec. 10, 1849. Introduction. vol. 2, part 1, pp. 1–26 [descriptions of the vegetation]

General geological features of the country. vol. 1, part 2, pp. 1–24 [authorship not attributed]

Geological features of the Rio Grande Valley from El Paso to the mouth of the Pecos River. vol. 1, part 2, pp. 49–61

U.S. Department of Agriculture. Reports

Character of herbarium, etc. 1869, p. 91

Royal Gardens at Kew, etc. 1870, p. 108

Botanical exploration in east Tennessee. 1871, p. 221(?)

Cinchona planting in Jamaica. 1872(?)

Utah Pomologist

The valley of the Virgen in 1844 and 1874. May, 1874

West American Scientist

New genus of Euphorbiaceae from Lower California. vol. 1, p. 13

Notes on *Chorizanthe lastarriaea.* loc. cit., no. 5

Historical notice of *Pinus torreyana.* loc. cit., no. 6

A new species of *Eriogonum* from Lower California. vol. 6, pp. 102–103

A handsome *Astragalus.* vol. 7, pp. 9–10

Western Weekly (Davenport)

Wild plants for house culture (essay). Feb. 21, 1874(read before the Scott County Horticultural Society)

References Cited

Arps, L. W., and E. E. Kingery. 1972. *High Country Names—Rocky Mountain National Park.* Johnson Publ. Co., Boulder, CO.

Dupree, A. Hunter. 1959. *Asa Gray, 1810–1888.* 505 pp. Harvard University Press, Cambridge, MA.

Engelmann, George. 1868. Altitude of Pike's Peak and other points in Colorado Territory. Trans. Acad. Sci. St. Louis 2:126–133.

Ewan, Joseph A. 1950. *Rocky Mountain Naturalists.* 358 pp., 9 portraits. Univ. Denver Press, Denver, CO.

———, and Nesta Ewan. 1981. Biographical dictionary of Rocky Mountain naturalists. Regnum Vegetabile 107:1–253.

Fairchild, David. 1939. *The World Was My Garden.* 494 pp. Charles Scribner's Sons, New York.

Farquhar, Francis Peloubet. 1961. Naming America's Mountains: The Colorado Rockies. American Alpine Journal 12:319–346.

Fisher, John. 1939. *A Builder of the West: The Life of William Jackson Palmer.* 332 pp. Caxton Printers, Caldwell, ID.

Flora of North America North of Mexico. 1993. Vol. 1. Introduction. 372 pp. FNA Editorial Committee. Oxford University Press.

Gabel, Mark L. 1981. The Parry Herbarium. Proc. Iowa Acad. Sci. 88:179.

Gannett, Henry. 1906. *A Gazetteer of Colorado.* USGS Bull. 291. Ser. F, Geography, No. 51.

Garnock-Jones, P. J., and C. J. Webb. 1996. The requirement to cite authors of plant names in botanical journals. TAXON 45: 285–286.

Goodman, George J., and Cheryl A. Lawson. 1995. *Retracing Major Stephen H. Longs's 1820 Expedition: The Itinerary and Botany.* 366 pp. Univ. Oklahoma Press, Norman, OK.

Gray, Asa. 1862–1863. Enumeration of the plants: by A. Gray, aided by the notes of Drs. Engelmann and Torrey, and upon the habitats, etc., by Dr. Parry. Amer. J. Sci. 33:237–243, 404–411; 34:249–261; 330–332.

———. 1863. Enumeration of the species of plants collected by Dr. C. C. Parry and Messrs. Elihu Hall and J. P. Harbour, during the summer and autumn of 1862, on and near the Rocky Mountains, in Colorado Territory, lat. 39°–41°. Proc. Acad. Nat. Sci. Philadelphia 15:55–80.

———. 1872. Dismissal of the late botanist of the Department of Agriculture. Amer. Nat. 6:3–7.

Harrington, H. D. 1954. *Manual of the Plants of Colorado.* 666 pp. Sage Books, Denver, CO.

Lenz, Lee W. 1986. *Marcus E. Jones, Western Geologist, Mining Engineer, and Botanist.* 486 pp. Rancho Santa Ana Bot. Garden, Claremont, CA.

McGregor, R. L., et al. 1977. *Atlas of the Flora of the Great Plains.* 600 pp. Iowa State Univ. Press, Ames, IA.

McKelvey, Susan Delano. 1955. *Botanical Exploration of the Trans-Mississippi West, 1790–1850.* 1144 pp., 11 maps. Arnold Arboretum of Harvard University, Jamaica Plain, MA.

McMullen, Glenn. 1995. Inventory of the Charles Christopher Parry Papers, 1841–1943. 4 pp. Manuscript Number MS-290. Dept. of Special Collections, The Parks Library, Iowa State University. Ames, IA.

McVaugh, Rogers. 1956. *Edward Palmer, Plant Explorer of the American West.* 430 pp. Univ. of Oklahoma Press, Norman, OK.

Mohlenbrock, Robert H. 1995. High Creek Fen, Colorado. Natural History 104(6):16–19.

Parry, C. C. 1862. Physiographic sketch of that portion of the Rocky Mountain range, at the head waters of South Clear Creek, and east of Middle Park: with an enumeration of the plants collected in this district, in the summer months of 1861. Amer. J. Sci. 33:231–243.

———. 1867. Notice of some additional observations on the physiography of the Rocky Mountains, made during the summer of 1864. Trans. Acad. Sci. St. Louis 2:272–286.

———. 1870. Botany of the region along the route of the Kansas Pacific Railway, through Kansas, Colorado, New Mexico, Arizona, and California. Appendix A, pp. 521–533 in W. A. Bell, *New Tracks in North America,* etc., ed. 2, Chapman & Hall, London.

———. 1876. Summer botanizing in the Wasatch Mountains, Utah Territory. Proc. Davenport Acad. Sci. 1:145–152.

Parry, C. C., and George Engelmann. 1862. Supplement to the enumeration of plants of Dr. Parry's collections in the Rocky Mountains, I. Coniferae. Amer. J. Sci. 34:330–341.

Parry, Mrs. E. R. 1891. Catalogue of the herbarium of the late Dr. Charles C. Parry. title page + 1. Printed by H. N. Patterson, Oquawka, IL.

———. 1893. List of papers published by the late Dr. C. C. Parry. Proc. Davenport Acad. Sci. 6:46–52. Reprinted from Bull. Torr. Bot. Club 24:322–328.

Preston, C. H. 1893. Biographical sketch of Dr. C. C. Parry. Proc. Davenport Acad. Sci. 6:35–45.

Rodgers, Andrew Denny, III. 1942. *John Torrey, A Story of North American Botany.* 352 pp. Princeton Univ. Press, Princeton, NJ. [This source contains much interesting detail concerning Parry's Colorado trips and Torrey's aid in appointing Parry to the Mexican Boundary Survey.]

Rydberg, Per Axel. 1906. *Flora of Colorado.* Colo. Agric. College, Agric. Exp. Sta. Bull. 100. 448 pp.

———. 1932. *Flora of the Prairies and Plains of Central North America.* 969 pp. New York Botanical Garden, New York.

Taylor, Bayard. 1867. *Colorado: A Summer Trip.* 185 pp. Reprinted 1989, University Press of Colorado, Niwot, CO.

Tuckerman, Edward. 1864a. Observationes lichenologicae [No. 3]: Observations on North American and other lichens. Proc. Amer. Acad. Arts Sci. 6:263–287.

———. 1866. *Lichens of California, Oregon, and the Rocky Mountains; so far as yet known.* 35 pp. J. S. & C. Adams, Amherst, MA.

———. 1882, 1888. *A Synopsis of the North American Lichens,* Part I, II. 261, 281 pp. S. E. Cassino, Boston, MA.

Weber, William A. 1958. Rediscovery of *Neoparrya.* Rhodora 60:265–271.

———. 1972. The Torrey Botanical Club centennial celebration of the dedication of Gray's and Torrey's Peaks. Souvenir booklet. 22 pp., map. Privately printed. With facsimile pages of *The Colorado Miner,* pp. 6–14. Thursday, Aug. 22, 1872. Georgetown, CO.

———, and Ronald C. Wittmann. 1992. *Catalog of the Colorado Flora.* 215 pp. University Press of Colorado, Niwot.

Index